T0239012

Springer Aerospace Technology

Series Editors

Sergio De Rosa, DII, University of Naples Federico II, Napoli, Italy

Yao Zheng, School of Aeronautics and Astronautics, Zhejiang University, Hangzhou, Zhejiang, China

Elena Popova, AirNavigation Bridge Russia, Chelyabinsk, Russia

The series explores the technology and the science related to the aircraft and spacecraft including concept, design, assembly, control and maintenance. The topics cover aircraft, missiles, space vehicles, aircraft engines and propulsion units. The volumes of the series present the fundamentals, the applications and the advances in all the fields related to aerospace engineering, including:

- structural analysis,
- aerodynamics,
- aeroelasticity,
- aeroacoustics,
- flight mechanics and dynamics
- orbital maneuvers,
- avionics,
- systems design,
- materials technology,
- launch technology,
- payload and satellite technology,
- space industry, medicine and biology.

The series' scope includes monographs, professional books, advanced textbooks, as well as selected contributions from specialized conferences and workshops.

The volumes of the series are single-blind peer-reviewed.

To submit a proposal or request further information, please contact: Mr. Pierpaolo Riva at pierpaolo.riva@springer.com (Europe and Americas) Mr. Mengchu Huang at mengchu.huang@springer.com (China)

The series is indexed in Scopus and Compendex

Anatoly Ivanovich Kozlov ·
Alexander Ivanovich Logvin ·
Oksana Gennadyevna Feoktistova ·
Dmitry Alexandrovich Zatuchny ·
Yuri Grigoryevich Shatrakov

Ice Structures for Airfield Construction

 Springer

Anatoly Ivanovich Kozlov
Moscow State Technical University of Civil
Aviation
Moscow, Russia

Alexander Ivanovich Logvin
Moscow State Technical University of Civil
Aviation
Moscow, Russia

Oksana Gennadyevna Feoktistova
Moscow State Technical University of Civil
Aviation
Moscow, Russia

Dmitry Alexandrovich Zatuchny
Moscow State Technical University of Civil
Aviation
Moscow, Russia

Yuri Grigoryevich Shatrakov
St. Petersburg State University
of Aerospace Instrumentation
St. Petersburg, Russia

ISSN 1869-1730   ISSN 1869-1749 (electronic)
Springer Aerospace Technology
ISBN 978-981-19-6213-4   ISBN 978-981-19-6211-0 (eBook)
https://doi.org/10.1007/978-981-19-6211-0

This Springer imprint is published by the registered company Springer Nature Singapore Pte Ltd.
The registered company address is: 152 Beach Road, #21-01/04 Gateway East, Singapore 189721,
Singapore

Introduction

The use of the ice cover of the Arctic zone and Antarctica for airfields (landing sites for aircraft) or as a location for radio navigation and radar facilities is an important and urgent task. This is primarily due to the large extent of these regions and the absence of another underlying surface. It should be noted that the use of various ices to solve this problem requires careful research.

The purpose of this book is to analyze various properties and structures of ice from the point of view of use in solving problems of civil aviation.

In recent years, various methods of studying the environment have been intensively developing all over the world, among which remote sensing methods have become the most important. The development of these methods is stimulated by the increasingly deteriorating environmental situation in the world, the need to solve various geological, geophysical, hydrophysical, and other problems. In addition, the tasks of studying forests, agricultural lands, the surface of the World Ocean, rivers and lakes, mountain ranges, etc. are of great importance. It is very important to study on a global scale the surfaces, covered with sea ice, i.e., the study of the state of these ice masses, their transformation and movement, since their influence on the global climate is well known. However, in addition to global observations, various local observations of sea ice are also important for solving a wide variety of tasks. These may include the choice of landing sites for aircraft for aviation, the determination of long-term wintering sites, the assessment of the possibility of conducting ships, using icebreakers, and others.

Global observations of large ice massifs can only be carried out by remote sensing methods, in particular from space, using artificial Earth satellites (AES). Local observations are also carried out by remote sensing methods, but can also be carried out from an aircraft (airplane or helicopter) or from any special platforms and with the help of sledge-tractor trains.

When carrying out remote sensing of sea ice, as well as any other surfaces, traditional problems arise to ensure maximum reliability of the obtained results, an adequate assessment of the state of sea ice, determination of its physico-mechanical, chemical, and other properties. Their solution requires the use of new remote sensing methods, improving the quality of processing received information, developing new algorithms for the identification and classification of sea ice.

At this time, many publications have appeared on the issues of remote sensing of sea ice, which are widely scattered in various journals and scientific papers and have not yet found proper reflection in the monographic literature.

It should be noted that the monograph also reflected some issues that were not related to remote sensing of sea ice, but related to the study of other surfaces. First of all, this applies to the description of various algorithms for the identification and classification of surfaces. Since the described algorithms can be successfully used in the identification and classification of sea ice, these publications are reflected in this monograph. At the same time, many methods of remote sensing of sea ice, described in this monograph, can be used for remote sensing of other types of surface, such as forests, swamps, and agricultural lands.

Contents

Abbreviations

AES   Artificial Earth satellites
CBB   A completely black body
EMW   Electromagnetic wave
IAHR   International Association for Hydraulic Research
IR   Infrared radiation
LFM   Linear frequency modulation
RS   Remote sensing
RSA   Radars with synthetic aperture
RSS   The remote sensing system
SVR   Side-view radars
WMO   Worldwide Meteorological Organization
UFR   The ultra high frequency range

Chapter 1
Modern Methods of Remote Exploration of Sea Ice for the Construction of Ice Airfields and the Location of Ground-Based Flight Support Facilities

1.1 Tasks and General Principles of Application of Remote Methods of Studying the Environment for Solving Problems of Civil Aviation

The use of the land surface, as well as seas and oceans, covered with ice, is currently used in civil aviation, as a rule, in two directions. Firstly, the ice surface is used as a means for landing various aircraft or for landing search parties, or for dumping valuable cargo. Secondly, ice must also be used for the location of flight support facilities. As a rule, such funds are placed in trailers that are artificially frozen in ice (Fig. 1.1).

It should be noted that for the construction of runways in a particular region, it is necessary to know not only the thickness of the ice but also its structure. Let us give an example. At a certain angle of inclination of the sun's rays, certain ice surfaces (for example, smooth) begin to melt. This leads to the formation of puddles on the runway, which makes it difficult to use it [1–3].

The concept of «remote methods of studying the environment» combines a variety of methods and tools that solve the task. Among the most widely used remote sensing methods, we note the following: acoustic, optical, active radar, passive radar, and complex. All of these methods are used to study the environment as a whole. Each of these methods has certain advantages and disadvantages. There is no universal method that could replace any other in the future. For example, the high information capabilities of optical methods, especially with the use of laser systems, are significantly limited by weather conditions. The presence of rain, fog, snow, etc. in certain cases practically does not allow measurements to be carried out by optical methods. Active radar methods are effective in cases, where the medium under study has high reflective properties. If the probed medium has low reflective properties, then the effectiveness of active radar measurements decreases. Below we consider the use of various methods of remote sensing of the environment in order to study sea ice [4].

When studying the environment by remote methods, a large complex of various tasks is solved. This includes the tasks of ecological control of the environment,

A. I. Kozlov et al., *Ice Structures for Airfield Construction*, Springer Aerospace Technology, https://doi.org/10.1007/978-981-19-6211-0_1

Fig. 1.1 The location of flight support facilities in Antarctica

i.e. finding ecologically unfavorable areas, determining oil spills and spots of other origins in the seas and oceans, and determining the degree of pollution of the atmosphere and land. Another set of tasks is related to determining the state of the probed object, for example, determining the speed and direction of wind over the ocean surface, which is extremely important for confident prediction of the occurrence of a storm; determination of the thickness and mass of the snow cover, which is very important for predicting the future melting of the snow cover and the corresponding flooding of rivers, as well as for predicting the future harvest; determination of the degree of maturity of crops over large areas in order to determine the optimal harvest time.

In addition to solving the tasks, listed above, remote methods make it possible to determine soil moisture, the degree of stress of the ice cover, for example, to assess the possibility of aircraft landing on this surface, the temperature profile of the atmosphere for weather forecasting, etc. [5, 6].

Any remote sensing system is a measuring system that is located on a selected carrier, for example, on an aircraft (airplane, helicopter, or on an artificial Earth satellite). Therefore, a certain value of the selected measured parameter, obtained in the measuring system (for example, with active radar surveillance, it can be the amplitude of the reflected signal, its frequency or phase, etc.) should be identified to the area, relative to which the carrier of the remote sensing system (RSS) is moving.

Then the reliability of the obtained data largely depends on the accuracy of the aircraft's location [7]. The total error of the aircraft location contributes to the error of the determined parameter of the underlying surface. Consequently, an increase in the efficiency of using RSS should be associated simultaneously with an increase in the quality indicators of the RSS carrier navigation systems, since incorrectly performed binding of the aircraft to the terrain can negate any accuracy of the measuring system.

In addition, it should be noted that in practice, in many cases, the task of transmitting measurement information from the aircraft to the Ground to ensure the operational processing of this information arises. Processing of the received information about the underlying surface (for example, the construction of radar images) due to the large volume of this information is not always possible directly on board the aircraft, and the requirement for the promptness of obtaining this information causes the need the use of powerful computing facilities, located in special processing points. Therefore, increasing the efficiency of obtaining data on the underlying surface requires the use of the on-board-Ground information channel, which also introduces a corresponding error in the value of the measured parameter of the underlying surface [4, 8].

Thus, in the general case of the variance of the error of the selected information parameter (parameters) consists of the sum of the error variances of the measuring system, the RSS carrier location system, and the data transmission information system, i.e. [9]

$$\sigma_e^2 = \sum_{i=1}^{n} K_{i_m} * \sigma^2_{i_m} + \sum_{k=1}^{l} K_{k_c} * \sigma^2_{k_c} + \sum_{j=1}^{m} K_{j_v} * \sigma^2_{j_v}, \tag{1.1}$$

where σ_e^2—the total variance of the error of the selected information parameter; $\sum_{i=1}^{n} K_{i_m} * \sigma^2_{im}$—total variance of the error of measurement system; K_{i_m}—the normalizing multiplier for the measuring system; $\sum_{k=1}^{l} K_{k_c} * \sigma^2_{k_c}$—the total variance of the error of the navigation positioning system of the RSS carrier; K_{k_c}—the normalizing multiplier for the navigation system; $\sum_{j=1}^{m} K_{j_v} * \sigma^2_{j_v}$—the total variance of the error of the system of transmission measuring information; K_{j_v}—the normalizing multiplier for the transmission measuring information, n, l, m,—the number of factors, determining the total variance of the error in the measuring system, the positioning system, and the data transmission system, respectively [3, 7, 10–13].

Equation (1.1) is written under the assumption that the errors of the listed systems are mutually independent, which is usually done in practice. Accordingly, it is advisable to choose the minimum variance of the error of the measured parameters as the criterion of RSS effectiveness, taking into account its intended purpose.

Thus, the RSS construction should be carried out in the direction, that all components of the RSS subsystem provide a minimum of variance of their errors, i.e. the measurement system minimizes the variance of the error in the allocation of an information parameter, the value of which determines the geophysical, physical, mechanical, or any other parameter of the underlying surface. The variance of the error of the parameters, determining the location of the aircraft, is minimized in the navigation system. In the data transmission system, the variance of the error in determining the values of information symbols is minimized. Obviously, these tasks require a comprehensive solution, since the optimization of one of the subsystems may not optimize the system as a whole.

However, assuming, that these RSS subsystems are mutually independent, it is possible to optimize each of the subsystems separately. Let us consider some aspects of the tasks, being solved to optimize the RSS subsystems, listed above.

In RSS measurement subsystems, first of all, the so-called inverse problems are solved, i.e. the characteristics and parameters of the reflected (if it is active radar) or radiated (if it is passive radar) signals determine the characteristics of the underlying surface. Such characteristics may include salinity of sea ice, volume of liquid phase in the ice cover, soil moisture, salt content in arable land, etc. It is obvious, that in order to obtain reliable information about the underlying surface, it is necessary to identify the corresponding parameters of the received signals as accurately as possible and have equations, linking the parameters of the received signals with the parameters of the underlying surface. The maximum accuracy of the selected parameters can be achieved by appropriate optimal processing of the received signals [14].

Obtaining the equations of dependence of the parameters of the received signals and various characteristics of the underlying surface in analytical terms is an almost unsolvable task due to its large multifactorial nature in most cases. Therefore, empirical dependencies, obtained on the basis of a large number of necessary experiments, are used in practice [12, 15]. Accordingly, there is a need to increase the information content of the measurements, i.e. obtaining the maximum possible information about the underlying surface by analyzing the structure and characteristics of the received signals. Accordingly, the more parameters of the received signal are used for processing, the more information about the state of the probed object is provided by the measurements.

To solve the problems of aircraft location, standard radio navigation equipment is used, not having sufficient accuracy in most cases to solve remote sensing problems. Here, certain hopes are pinned on the creation of satellite radio navigation systems, the accuracy of which fully meets the needs of RSS. At the same time, we note, that RSS itself can solve navigation and location problems. In the presence of certain landmarks (of natural or artificial origin), RSS performs the functions of a navigation system. In this case, it is necessary to have in advance an appropriate map (in the form of radar, television, optical, or other images) of the terrain, stored in the memory of the on-board computer and use this map to verify the location of the aircraft in relation to this map [16–19].

Finally, the requirements for the data transmission subsystem are as follows. This RSS subsystem must ensure the necessary efficiency of data transmission, i.e. data processing must be carried out in real time, ensure the reliability of transmission, i.e. maintain a given probability of error in the reception of one information symbol. For this purpose, special types of signals and optimal methods of their processing should be used in the RSS data transmission system.

Of these RSS subsystems, only RSS measurement subsystems are considered in this monograph, since the construction of navigation systems and information transmission systems for various purposes and, in particular, for RSS is widely covered in the corresponding literature.

The tasks, listed above, solved with the help of RSS, relate in general to the study of the environment, but further attention will be paid to remote methods of studying sea ice. Note, that the actual remote sensing methods, described in relation to the study of sea ice are naturally suitable for remote sensing of any other surface. Many models, in which ice sheets are described, are also valid for other surfaces. In some cases, appropriate explanations will be given in the text. At the same time, some well-known methods of surface identification, recently proposed for specific surfaces, for example, for the identification of mixed forests, can be methodologically applicable for the recognition of age gradations of sea ice.

Further, for the general case, we will give a brief description of various remote sensing methods, the detailed description of which is given appropriate attention in this book in relation to the study of sea ice.

1.2 Acoustic Methods

The essence of the acoustic method and, in particular, the most widely used hydroacoustic method consists in the presence of a certain dependence between the energy and spectral characteristics of acoustic signals, propagating in an aqueous medium or reflected from its boundaries with the physical characteristics of this medium. Remote acoustic methods, by analyzing the angular, frequency, and energy characteristics of sound signals, reflected from the probed surface, make it possible to find certain physical parameters of the surface and determine its geometric abilities.

For the application of the hydroacoustic method of remote sensing (RS), certain models of surfaces are constructed, taking into account, first of all, their roughness. Usually, two types of models are considered in hydroacoustics: fine-grained and coarse-grained surfaces. In the first case, the method of small perturbations is used to analyze the processes of acoustic wave scattering, and in the second—the Kirchhoff method.

In the general case of reflection, an acoustic wave is a certain total field, containing coherent and incoherent components [20]. In this case, the coherent component obeys the laws of geometric optics, and the non-coherent component contains random amplitude and phase. In various hydroacoustic devices, either one or the other component of the total field is used, or both together.

The existing relationships between the characteristics of the scattered acoustic field and the parameters, describing this medium, are very difficult to imagine in an analytical form. In addition, such connections should be probabilistic in nature, taking into account the statistical nature of the roughness of the surface. However, there are quite a few relevant statistical data, which is why researchers usually just use the Gaussian distribution of the heights of irregularities on the surface. However, in many real cases, such a simplified approach is incorrect and does not reflect the actual picture. Therefore, basic information about the characteristics of the sound wave, scattering on various surfaces, is obtained from laboratory and field experiments [5, 8, 21–25].

It should be noted, that with the help of hydroacoustic methods, it is possible to solve various problems of remote sensing of the studied surfaces. Let us list some of them with a brief description of their capabilities.

One of the methods of hydroacoustic RS is the method of hydroacoustic profiling of the lower surface of the coverage. This method is based on the presence of a relationship between the energy and spectral characteristics of the reflected acoustic wave signals and the physical parameters of the probed object, and it is assumed, that the sound propagates rectilinearly at small angles of incidence. The spatial position of the probed object is determined by finding the range, angle of location, and azimuth.

The hydroacoustic profiling method makes it possible to solve the problems of constructing the relief of a probed object, only in a very narrow band along the route of the receiving antenna carrier, but it does not provide information about the areal changes in its irregularities [26]. To solve this problem, the method of mapping the relief of the lower surface of the object, using side-view sonars, is used. In this case, it becomes possible to obtain two-dimensional sonar images of sound-scattering surfaces. To construct two-dimensional sonar images, spatio-temporal signal, processing with phased antenna arrays, is used [27].

The above hydroacoustic methods give a statistical picture of the probed surface. At the same time, it is very important to determine the dynamics of changes in the cover, when studying various covers. For this purpose, an acoustic method is used to study variability, for example, the thickness of the ice sheet. In this method, the fix of change in the position of the lower boundary of the ice sheet relative to some initial level is performed. Then, with a certain discreteness in time, the distance between the source of the audio signal and the lower boundary of the ice is fixed.

In addition to the hydroacoustic method for determining the dynamics of the surface, its modification is used—the radiohydroacoustic method. The essence of the method is that a hydroacoustic and radio channel are used together, and a radio signal receiver is located at the location of the audio signal transmitter and a radio signal transmitter is located at the location of the acoustic signal receiver [28–30]. With tight synchronization between the reception of an acoustic signal and the transmission of a radio signal, it is possible to obtain high accuracy in determining the change in the distance between two observation points with consecutive measurements in time.

Another area of application of hydroacoustic remote sensing methods is the measurement of the drift velocity vector of the probed object. Such problems arise, when determining the rate of drift of ice or oil slicks on the ocean surface. In this case, one emitter and two acoustic receivers are used. The measurement of the velocity and direction of drift of the probed object is based on the measurement of the phase shift between the radiated oscillations and the oscillations, arriving at the receiving point.

The advantages of hydroacoustic remote sensing methods include a relatively high accuracy of determining the specified parameters, the possibility of probing the object as, if from the inside, as well as the determination of various properties of the object in a limited area, which gives a more accurate picture of the state of the object. At the same time, hydroacoustic methods have limitations in their application. These methods, for example, do not allow conducting research on large-scale objects with sufficient accuracy and are ineffective in external measurements.

1.3 Optical Methods

The optical methods of remote sensing include aerial photography, satellite television photography and the use of various laser systems [31].

Aerial photography allows you to solve many problems, related to the observation of large surface areas, for example, determining the boundaries of changes in the type of surface (swamps, forests, water surface, etc.), determining the boundaries of changes in the type of vegetation of a given surface (coniferous forest, deciduous, mixed, etc.), finding faults, depressions, cracks, and other violations of the terrestrial the cover. However, the solution to these tasks in aerial photography is possible only with the help of non-traditional methods of aerial photography, for example, when using a circular planned-perspective aerial camera, in which the image in the focal plane is projected not through one, but through two lenses, the optical axes of which are parallel. Such a system ensures the capture of the entire visible plane and gives high resolution and uniform illumination over the entire field of the frame.

A significant expansion of the areas, covered directly by observation is achieved by using satellite photography. It should be noted that simultaneously with satellite photography, satellite television surveys are widely used, which are especially effective in determining the dynamics of the behavior of the studied surface. But this efficiency can be achieved only, if the maximum possible accuracy of the geographical reference of the received data is obtained, which requires certain additional efforts [30, 32]. Thus, aerial photography and television photography can provide such visual information about the probed surfaces, that other known remote sensing methods cannot provide. However, the methods of aerial photography and television shooting have a strong dependence on weather conditions. Television shooting largely depends on the illumination of the surface and therefore can be used only in a limited way, only at certain times of the year. In addition, the presence of cloud cover over the probed surface during filming makes it practically impossible to conduct them [33]. There are other factors that make it difficult to conduct aerial photography and television shooting, described in more detail below. All this points to the limited use of optical remote sensing devices, which have recently been mainly used in combination with other remote sensing methods, for example, radar. In this case, television images of the surface are compared with radar images. Separately, it should be pointed out, that remote sensing can be carried out, using optical means, such as laser systems, that allow profiling of the upper surface of the Earth's covers. Moreover, potentially such profiling can be performed with an accuracy, unreachable for any other remote sensing systems. However, achieving these potential accuracies is associated with a number of significant difficulties, overcoming which is a complex scientific and technical task. Such difficulties include: the need to strictly maintain a certain position, relative to the RSS (for example, the aircraft), since changing the position of the aircraft in height during measurements affects the accuracy of laser profiling of the surface; the need to use complex systems for processing received information; the strong influence of weather conditions on measurements, using laser means [34–36].

The same difficulties are encountered when using laser technology to study the dynamics of the probed object. RS laser systems use laser interferometers, which make it possible to achieve high accuracy of measurements of the deformation of the probed surface and a high degree of resolution. At the same time, an additional difficulty arises due to the influence of the instability of the refractive index of the atmosphere on the measurement process.

Thus, optical remote sensing methods, including laser, having high potential capabilities, have many limitations in terms of actual application and in most cases should be used in combination with other remote sensing tools.

1.4 Active Radar Methods

With active radar sensing, an onboard radar station is used, which emits a probing signal and receives the necessary information by measuring the parameters of the reflected radar signal. There may be various modifications to these radar facilities. These include a radio altimeter that determines the altitude of the aircraft over the underlying surface and thereby allows you to build a relief of the probed surface; a scatterometer determines the conditions and characteristics of the scattering of radio signals from the probed surface [37]. At the same time, radar facilities can be of various types: radars with synthetic aperture (RSA), side-view radars (SVR), etc.

These radar facilities can operate in different frequency ranges, while in some cases, for different surfaces, a change in the frequency range may not affect the characteristics of the received signals, and in others, significant changes in these characteristics are observed. With active remote sensing, two fundamental problems are solved: the direct problem, i.e. determining the characteristics of the received signal, depending on the type and condition of the underlying surface; the inverse problem, when the type and condition of the underlying surface are determined by the characteristics of the received signal. Obviously, to solve the problems, formulated above, the solution of inverse problems is most interesting. However, it should be borne in mind, that the solution of inverse problems (usually quite complex) can often rely on the solution of known direct problems [38, 39].

Active radar means of sensing the surface can be classified in several directions. The main directions of such classification are by purpose, by the type of surface under study, by the location of the radar system (radar), by the principle of station construction, and by a number of other features (frequency range, type of polarization of the emitted signal, type of used probing signals, etc.).

The most complete characteristic of the probed surface is its scattering matrix, obtained for various frequencies and viewing angles. The elements of the scattering matrix are determined by amplitude-phase measurements. At the same time, it should be borne in mind, that with absolute phase measurements it is impossible to achieve high accuracy due to the incommensurability of the distance between the probing radar and the object and the wavelength [35, 37, 40–42]. It makes real sense to measure the phase difference between the orthogonal polarization components, which is sufficient to determine the division of the parameters for the scattering matrix.

For remote sensing purposes, pulse signals are most often used, although radars with a continuous frequency-modulated signal are used to probe, for example, ice sheets at a carrier frequency equal to 400 MHz or lower. In addition, modern radars use complex signals with linear frequency modulation (LFM) and phase-manipulated signals, built, for example, on the basis of Barker codes, to carry out penetrating sensing. The complexity of the tasks of recognizing the physical parameters of the probed object lies in the need to calibrate the radar and determine the specific effective scattering surface σ^0, measured, using the mentioned scatterometers (σ^0 is often also called the backscattering coefficient).

It should be noted, that the scattering properties of a distributed target are fully determined, if the dependences of the backscattering coefficients σ^0 on the viewing angle Θ, frequency f, and the type of polarization of the probing electromagnetic wave (EMW) are known. In turn, for a given observation angle, frequency, and type of EMV polarization, the backscattering coefficient σ^0 is determined by the dielectric and geometric properties of the probed object. Therefore, having the values of σ^0 and having the necessary functional connections between the changes of σ^0 and the dielectric properties of the probed object (using certain models of surface roughness), it is possible to determine the real physical, chemical, and other properties of the underlying surface, which are determined by its electrical properties. It follows from this, that in order to solve RS problems, it is necessary to use RSA, operating at different viewing angles in the conditions of the multi-frequency mode of the emitted signal and with different types of EMV polarization [43]. Therefore, modern radars of RS operate at different sensing frequencies and have the ability to choose different types of polarization, emitted by EMV.

In addition to the main type of EMV polarization, emitted and received in the radar receiver, additional information about the object, being probed, is also carried by the cross-polarization component of backscattering, which occurs due to the effect of volumetric scattering on inhomogeneities of this volume. The use of this effect makes it possible to increase the information content of the process of recognizing the properties of an object, however, its presence leads to a decrease in the signal level, received at the main EMV polarization, and the reception and processing of the cross-polarization component entail a complication of the system as a whole. Nevertheless, some properties of the probed objects can be revealed only by taking into account the cross component of the received electromagnetic radiation [37, 38]. Therefore, the methods of analyzing the polarization state of radio waves, reflected from the probed objects are widely used in modern RSS and currently constitute a whole field of research, called radio polarimetry.

1.5 Passive Radar (Radiometric) Methods

Passive radar sensing uses an onboard radiometer that receives and processes signals of its own radiothermal radiation from the underlying surface of the atmosphere. There are various modifications of radiometers, related to the features of their construction. Radar sensing devices, both active and passive, can operate in various frequency ranges, up to 200 GHz frequencies. The physical essence of the passive radar sensing method is the effect of generating fluctuating electromagnetic fields by random chaotic currents of certain volumes, as a result of which any real physical body creates natural radiothermal radiation, caused by electrodynamic processes, occurring in its atoms and molecules [44]. The physical essence of this process is the transformation of the internal thermal energy of the body into the energy of the electromagnetic field. The transformation is carried out by a set of elementary oscillators, excited by the thermal motion of particles of matter in a state of chaotic thermal motion and transferring part of their kinetic energy to oscillators. This process is accompanied by electromagnetic radiation, containing a base for the implementation of passive radar and the construction of radiometers. At the same time, the intensity of its own radiation in the microwave range is characterized by the radio brightness temperature T_b, determined by the product of the radiation coefficient x of the medium by its effective temperature T_0. The radiation coefficient of a medium, homogeneous in depth, bounded by a flat plane, at the length of the electromagnetic wave λ is described, using Fresnel reflection formulas, which include the characteristics of the medium in the form of a complex permittivity $\varepsilon_\lambda = \varepsilon'_\lambda - j\varepsilon''_\lambda$, the viewing angles θ, and are determined by the polarization of the radiation—vertical or horizontal [43–45]. The sum of the emission coefficient, in $(\varepsilon_\lambda; ; \theta)$ and the reflection coefficient in $(\varepsilon_\lambda; \theta)$ is equal to one in the first approximation, if we do not take into account a number of additional effects, that occur when the EVM passes through the interface of the medium.

Hence, the physical meaning of radiometric analysis is to measure the radio brightness temperatures T_b at different polarizations of radiation at different viewing angles θ, which makes it possible to associate changes of T_b with the electrophysical characteristics of the medium, i.e. with ε_λ, changes in which are associated, for example, with soil moisture, with the salinity of sea ice, its age characteristics, with the volume of the liquid phase, etc. Note that the underlying surface (the relief of the surface, the hummockiness of the ice cover) plays an important role in measuring T_b and taking into account the influence of radiothermal radiation of the atmosphere, entering the receiving device of the radiometer.

The equipment for receiving and measuring flows of radiated energy in the corresponding ranges of the electromagnetic spectrum should have the following properties [46]:

– receive radiation with a certain spatial resolution;
– have high sensitivity;
– quantify radiation flows with high accuracy;

– to provide the possibility of unambiguous binding of the measured flows to the spatial coordinates of the corresponding radiating elements.

These requirements determine the construction of radiometric equipment as a set of three devices: an antenna system for converting heat flow into antenna temperature; a radiometric receiver, that allows measuring antenna temperature; a display indicator, that allows comparing the positions of the emitting element in space with the intensity of its radiation [47].

Existing radiometers are usually multichannel (frequency diversity), scanning (angle changes θ), with varying polarization of receiving antennas (in each frequency channel). Such radiometric systems are used to determine the dynamics of global ice surfaces, as well as for local radiation of selected areas of ice coverings. As in active radar, with the help of radiometers, the problems of identification and classification of the probed surfaces, in particular sea ice, are considered. A special role is assigned to radiometers in solving the problem of determining the age boundaries of sea ice.

1.6 The Method of Integration of Active and Passive Radar Sensing Means in Order to Use the Underlying Surface to Ensure Aircraft Flights

In recent years, in order to increase the information content and reliability of the obtained data, integrated remote sensing systems have been developed, combining active, and passive radar facilities and optical systems, based on the use of laser devices. A special case of such integration, but the most widely used in practice, is a complex remote radio sensing system, combining active and passive radar systems. In this case, the active radar and the radiometer, located on board the aircraft (or AES), simultaneously probe the same surface. As a result, the obtained measurement data complement each other. At the same time, two main directions are possible in the work of the complex [45, 48, 49]. In the first case, the active and passive systems work independently, probing the same object either with a time difference or simultaneously. Then the measurements of one of the systems serve to correct or refine the data, obtained by the other system. This is most often used when constructing radar images of the surface and highlighting its most characteristic features.

In the second case, simultaneous measurements are carried out in such a way, that during the measurement process there is a continuous correction of the data, obtained by comparing information, coming from two or more information sensors, which are used as active and passive radars [50].

Thus, for the first case of construction of complex RS systems, the aggregativeness of the system is characteristic, i.e. the usual unification of various systems, solving a single problem, occurs, without their internal unification. It is technically quite simple to perform such integration, therefore this type of integration has found wide application in practice and is most fully presented in relevant publications.

It is obvious, that the second case of aggregation requires a significantly more complex system for processing the received information. Therefore there are few such «truly» complex systems and their description is extremely insufficiently presented in the literature. However, with this method of integration, it is possible to realize to the maximum extent the possibilities, that the integration of various systems can give.

References

1. Eliseev BP, Kozlov AI, Romancheva NI, Shatrakov YG, Zatuchny DA, Zavalishin OI (2020) Probabilistic-statistical approaches to the prediction of aircraft navigation systems condition. Springer Aerospace Technology, p 200
2. Akinshin RN, Zatuchny DA, Shevchenko DV (2018) Reducing the effect of the multipath effect, when transmitting information from an aircraft. Informatiz Commun 3:6–12
3. Zatuchny DA (2018) Analysis of the impact of various interference on the navigation systems of civil aviation aircraft. Informatiz Commun 2:7–11
4. Breitkraits SG, Zatuchny DA, Ilyin EM, Polubekhin AI (2017) A methodical approach to determining the place of non-traditional radar methods in promising radar applications. Bull St Petersburg State Univ Civ Aviat 3(16):72–84
5. Zatuchny DA, Logvin AI (2012) Satellite navigation systems and air traffic management. Study guide. Publishing House of the Moscow State Technical University of Civil Aviation
6. Eliseev BP, Kozlov AI, Sarychev VA, Fadeeva VN (2013) Volume V. The theory of electromagnetism from electrostatics to radio polarimetry. Part 8. Electromagnetic waves. Radio Electronics, Moscow
7. Shatrakov YG (2015) Flight safety and the direction of development of simulators for air traffic management specialists. Publishing House of the State University of Aerospace Instrumentation, St. Petersburg, p 516
8. Kuklev EA, Shapkin VS, Filippov VL, Shatrakov YG (2020) Aviation systems risks and safety. Springer Aerospace Technology
9. Kozlov AI, Logvin AI, Sarychev VA, Shatrakov YG, Zavalishin OI (2020) Introduction to the theory of radiopolarimetric navigation systems. Springer Aerospace Technology
10. Didenko NI, Eliseev BP, Sauta OI, Shatrakov AY, Yushkov AV (2017) Radio engineering support of military and civil aviation flights—a strategic problem of the Arctic zone of Russia. Sci Bull Mosc State Tech Univ Civ Aviat 20(5):8–19
11. Lipgart LP, Kozlov AI, Logvin AI, Vatin IV (2019) Polarization methods for determination and visualization of complex dielectric permittivity in remote sensing issues. Sci Bull Mosc State Tech Univ Civ Aviat 22(4):100–108
12. Kozlov AI, Sergeev VG (1998) Propagation of radio waves along natural routes. Moscow
13. Kozlov AI, Maslov VY (2004) Differential properties of the scattering matrix. Sci Bull Mosc State Tech Univ Civ Aviat 79:43–46
14. Shatrakov YG (2018) Development of domestic radar aircraft systems. Publishing House Stolichnaya Encyclopedia, Moscow, p 400
15. Drachev AN, Farafonov VG, Balashov VM (2014) Methods of control of complex profile surfaces. Radio Electron Issues 1(1):91–99
16. Antsev GV, Bondarenko AV, Golovachev MV, Kochetov AV, Lukashov KG, Mironov OS, Panfilov PS, Parusov VA, Raisky VL, Sarychev VA (2017) Radiophysical support of ultrashort pulse radar systems. In: Problems of remote sensing, propagation and diffraction of radio waves. Lecture notes. Scientific Council of the Russian Academy of Sciences on Radio Wave Propagation, Murom Institute (Branch), Vladimir State University, Named After Alexander Grigoryevich and Nikolai Grigoryevich Stoletov, pp 5–21

17. Antsev GV, Bondarenko AV, Golovachev MV, Kochetov AV, Mironov OS, Panfilov PS, Parusov VA, Sarychev VA (2016) Technologies of ultrashort pulse radar of natural environments with high range resolution. Meteorol Bull 8(3):17–22

18. Antsev GV, Bondarenko AV, Golovachev MV, Kochetov AV, Lukashov KG, Mironov OS, Panfilov PS, Parusov VA, Raisky VL, Sarychev VA (2016) Experimental studies of radar system characteristics. In: Radiophysical methods in remote sensing of medium. Materials of the VII All-Russian scientific conference. Murom Institute (Branch) Federal State Budgetary Educational Institution of Higher Education, Vladimir State University, Named After Alexander Grigoryevich and Nikolai Grigoryevich Stoletov, pp 196–202

19. Balashov VM, Drachev AN, Michurin SV (2019) Methods of control of reflectors of mirror antennas. In: Metrological support of innovative technologies. International Forum: Abstracts, pp 44–46

20. Balashov VM, Drachev AN, Smirnov AO (2019) Methods of coordinate measurements in the control of a complex profile surface. In: Metrological support of innovative technologies. International Forum: Abstracts, pp 41–43

21. Zatuchny DA (2017) Analysis of the features of wave reflection, when transmitting data from an aircraft in urban conditions. Informatiz Commun 2:7–9

22. Zatuchny DA, Kozlov AI, Trushin AV (2018) Distinguishing observation objects, located within the irradiated area of the surface. Informatiz Commun 5:12–21

23. Kozlov AI, Logvin AI, Sarychev VA (2007) Radar polarimetry. Polarization structure of radar signals. Radio Engineering, p 640

24. Proshin AA, Goryachev NV, Yurkov NK (2018) Calculation of radio wave attenuation. Certificate about registration of the computer program RUS 2019612561, 5 Dec 2018

25. Rassadin AE (2010) The apparatus of atomic functions and R-functions as the basis of the mathematical design technology of the RSA on an air carrier. In: Pupkov KA (ed) Intelligent systems: proceedings of the ninth international symposium. RUSAKI, Moscow, pp 224–228

26. Kozlov AI, Amninov EV, Varenitsa YI, Rumyantsev VL (2016) Polarimetric algorithms for detecting radar objects against the background of active noise interference. Proc Tula State Univ: Tech Sci 12-1:179–187

27. Sinitsyn VA, Sinitsyn EA, Strakhov SY, Matveev SA (2016) Methods of signal formation and processing in primary radar stations. St. Petersburg

28. Davydov PS (ed) (1984) Aviation radar: handbook. Transport, Moscow, p 223

29. Gromov GN, Ivanov YV, Savelyev TG, Sinitsyn EA (2002) Adaptive spatial Doppler processing of echo signals in radar systems of air traffic management. Federal State Unitary Enterprise «All-Russian Scientific Research Institute of Radio Equipment», St. Petersburg, p 270

30. Zhuravlev AK, Khlebnikov VA, Rodimov AP et al (1991) Adaptive radio engineering systems with antenna arrays. Publishing House of Leningrad State University, Leningrad, p 544

31. Kachurin LG, Divinsky LI (eds) (1992) Active-passive radar of thunderstorms and lightning-dangerous sites in clouds. Hydrometeorological Publishing House, St. Petersburg, p 216

32. Drogalin VV, Merkulov VI, Rodzivilov VA et al (1998) Algorithms for estimating angular coordinates of radiation sources, based on spectral analysis methods. Foreign Radio Electron: Successes Mod Radio Electron 2:3–17

33. Alekseev VG (2000) About nonparametric estimates of spectral density. Radio Eng Electron 45(2):185–190

34. Andreev VG, Koshelev VI, Loginov SN (2002) Algorithms and means of spectral analysis of signals with a large dynamic range. Radio Electron Issues. Ser Radar Equip (1–2):77–89

35. Bean BR, Dutton ED (1971) Radiometeorology. Hydrometeorological Publishing House, Leningrad, p 362

36. Brylev GB, Gashina SB, Nizdoiminoga GL (1986) Radar characteristics of clouds and precipitation. Hydrometeorological Publishing House, Leningrad, p 231

37. Van Tris G (1977) Theory of detection, evaluation and modulation, vol 3 (translated from English). Soviet Radio, Moscow, p 664

38. Vityazev VV (1993) Digital frequency selection of signals. Radio and Communications, Moscow, p 239

39. Vlasenko VA, Lappa YM, Yaroslavsky LP (1990) Methods of synthesis of fast convolution algorithms and spectral analysis of signals. Nauka, Moscow, p 180
40. Vostrenkov VM, Ivanov AA, Pinsky MB (1989) Application of adaptive filtration methods in Doppler meteorological radar. Meteorol Hydrol 10:114–119
41. Vostrenkov VM, Melnichuk YV (1984) Signals of the underlying surface and meteorological objects on the onboard Doppler radar. Proc Cent Aerolog Obs (154):52–65
42. Gritsunov AV (2003) The choice of methods for spectral estimation of time functions in the modeling of ultrahigh frequency devices. Radio Eng 9:25–30
43. Gun S, Rao D, Arun K (1989) Spectral analysis: from conventional methods to methods with high resolution. In: Gun S, Whitehouse H, Kailat T (eds) Ultra-large integrated circuits and modern signal processing (translated from English, Leksachenko VA (ed)). Radio and Communications, Moscow, pp 45–64
44. Jenkins G, Watts D (1971–1972) Spectral analysis and its applications: translated from English in 2 volumes. Mir, Moscow
45. Dorozhkin NS, Zhukov VY, Melnikov VM (1993) Doppler channels for radar MRL-5. Meteorol Hydrol 4:108–112
46. Ermolaev VT, Maltsev AA, Rodyushkin KV (2000) Statistical characteristics of AIC, MDL criteria in the problem of detecting multidimensional signals in the case of a short sample: report. In: The third international conference «digital signal processing and its application»: reports, vol 1. Moscow, pp 102–105
47. Zhuravlev AK, Lukoshkin AP, Poddubny SS (1983) Signal processing in adaptive antenna arrays. Publishing House of Leningrad State University, Leningrad, p 240
48. Zubkovich SG (1968) Statistical characteristics of radio signals, reflected from the Earth's surface. Soviet Radio, Moscow, p 224
49. Vereshchagin AV, Zatuchny DA, Sinitsyn VA, Sinitsyn EA, Shatrakov YG (2020) Signal processing of airborne radar stations plane flight control in difficult meteoconditions. Springer Aerospace Technology, p 218
50. Yurkov NK, Bukharov AY, Zatuchny DA (2021) Signal polarization selection for aircraft radar control—models and methods. Springer Aerospace Technology, p 140

Chapter 2
Structure and Composition of the Sea Ice Cover

The Arctic and Southern Oceans are unique regions of concentration of sea ice. Other seas, adjacent to the polar regions, also have significant masses of sea ice. The sea ice of the Arctic Ocean covers almost its entire surface, equal to 13,100,000 km^2, throughout the year. By the end of winter (April–May), sea ice covers an area of 10,800,000 km^2, and by the end of polar summer (August)—8,000,000 km^2. The presence of year-round ice cover is the most important characteristic feature of the Arctic basin. Sea ice has a complex morphological structure. Their thickness, the roughness of the upper and lower surfaces have a temporary variability. The physical properties of the ice cover also change significantly [1].

The difference in the morphological characteristics of ice sheets according to their main parameters—thickness, hummockiness, relief of the upper and lower surfaces is due to the action of a number of hydrometeorological factors and ice drift. Ice drift causes deformations that lead to the formation of cracks, the appearance of hummocks and cracks [2]. Thermal processes cause the formation and melting of ice: their seasonal variability contributes to the formation and further transformation of the relief of the ice cover. Wind and water erosion leads to the smoothing of the ice relief. Marine bioorganisms and water pollution enhance the course of local thermal erosion processes.

The impossibility of differentiating the role of various factors in the formation of the thickness of the ice cover and its relief leads to the fact that in each case their cumulative effect is observed. Therefore, statistical methods are used to study general laws, reflecting the state of the ice cover, while the measured parameters of the state of the ice cover are considered as random variables [3–5].

A. I. Kozlov et al., *Ice Structures for Airfield Construction*, Springer Aerospace Technology, https://doi.org/10.1007/978-981-19-6211-0_2

2.1   Types of Sea Ice and Nomenclature of Ice Terms of the World Meteorological Organization. The Possibility of Using Different Types of Ice Sheets to Support Aircraft Flights

The variety of conditions for the formation of the sea ice cover is the reason for the existence on the surface of the freezing waters of a large number of the most diverse types of ice in structure and physical properties.

The main terms, characterizing floating ice and related phenomena, have been used for a long time. However, due to the development of new, for example remote methods for studying sea ice, new terms, such as sonar, radar, and infrared images of the ice cover appear. Ice terminology has been standardized Worldwide Meteorological Organization (WMO) [2, 6–8]. The main terms that are found in this book are given according to the current WMO nomenclature. The age characteristics of ice include: age composition of ice, initial types of ice, nilas, pancake ice, young ice, annual ice, and old ice. Forms of fixed ice: soldered; soldered sole; bottom ice; and stranded ice. The drifting ice is characterized by its iciness, cohesion, forms of floating ice, and ice distribution [4, 9]. To describe the dynamics of drifting ice, the following terms are used: ice spreading, ice consolidation, ice movement, and ice drift. The following terms refer to the processes of ice deformation: ice breaking, ice layering, humping, and smoothing. When describing the space of water among the ice, the terms are used: spreading, spreading zone, channel, polynya, and washout. The characteristic of the ice surface can be given, using the terms: smooth ice, deformed ice, snow-covered ice, dirty ice, etc. The stages of melting are characterized by the destruction of ice, snowflakes, thawed ice, dried ice, rotten ice, and battering ram. Some characteristics of the main ice gradations are given in Table 2.1.

2.1.1   Formation and Development of the Ice Cover. Initial Ice Forms

The process of freezing of seawater differs significantly from the processes of crystallization of fresh water. Firstly, the water temperature, at which sea ice is formed is not a constant and depends on its salinity [10]. For example, for water with a salinity of 35‰, it is equal to 1.91 °C, 30‰—1.63 °C, etc. Secondly, when sea water freezes with a salinity of more than 24.7‰, a process of vertical convection takes place on the surface as a consequence of the fact, that, when cooling, the water becomes denser and sinks, replacing the underlying layers. In coastal Arctic regions, water mixing due to vertical convection is usually observed in the upper layer with a thickness of 10–20 m. In a freshwater reservoir, as it is known, denser water is at the bottom and its temperature is 4 °C, so the process of ice formation at a water temperature of about 0 °C goes without vertical convection [5, 11, 12].

Table 2.1 The increasing gradation of sea ice, their cohesion and some other characteristics

Age characteristics	Cohesion, scores (on a 10-point scale)	Size of ice fields	Classification by extent of drifting ice
Initial forms: ice needles ice fat snow shuga	Solid, 10 Frozen, 10 Very close-knit, 9–10	Gigantic (>10 km) Extensive (2–10 km) Fragments of ice fields (100–200 m)	The accumulation of deforming ice: large (>20 km) medium (15–20 km) small (10–15 km)
Young: gray (10–15 cm) gray-white (15–30 cm)	Clean water (without ice)	Grated ice (<2 m) Ice porridge—accumulation of grated ice	
Annual: thin (30–70 cm) medium (70–120 cm) thick (>120 cm)		Ice islands and icebergs are, as a rule, of glacial origin and are allocated to a separate class	
Old: two year long term			

Thus, as soon as the air temperature drops to –2 °C, small crystals of needle or plate ice form on the surface of the sea. These crystals often have a vertical orientation of the C-axes. This is especially the case, if the core of crystallization of supercooled water are snowflakes. A thin layer of snow forms on the surface. A thick viscous layer of ice needles, reflecting little light (the surface acquires a matte shade), is known as ice fat. The ice crust, which has hardened in calm conditions with a thickness of about 5 cm, easily breaks, when the sea surface is curved. In this case, discs of young ice of round or oval shape with a diameter of 0.3–3.0 m are formed. The discs have raised edges in the form of a whitish roller of worn ice. Such ice is called pancake ice [4, 7]. Its thickness is about 10 cm. The next age stage of the relatively stable state of the sea ice cover is *young ice*. These are *gray thin ice*, which easily bend on a wave, layer up in the wind and form a stable continuous layer on the surface of the water. The age of young ice varies from a few hours to 25–30 days, depending on the geographical location of the water area and the climatic conditions of a particular year. The thickness of young ice is 5–30 cm. Young ice is divided into nilas (dark and light), flask, gray ice and gray-white ice.

2.1.2 Annual Ice

In the primary layer of ice, formed on the sea surface under the influence of changing environmental conditions, various dynamic phenomena are usually first observed, leading to the destruction of a thin ice sheet and the layering of ice fragments on top of each other. Only under relatively stable conditions at the ice–water boundary the stable growth of crystals in the downward direction begin to prevail [13]. This process is slow: at its typical speed of about 1 cm per day, crystals of about 1–10 cm in size, and sometimes more, are formed. At the initial stage of growth, these crystals are observed in the form of thin plates, extending downward under the lower surface of a solid ice sheet; they are called *intrawater ice*. The crystals elongate in the vertical direction, forming a structure, consisting of plates approximately 1 mm thick with C-axes, directed horizontally.

The predominance of the horizontal orientation of the C-axes is due to the predominant growth of crystals, oriented along the vertical heat flow. In the same direction, i.e. perpendicular to the C-axes of the crystals, ice has a maximum thermal conductivity [14–16]. As fresh crystals form, salt is displaced into the space between the growing plates. At the ice–water boundary, cold brine streams flow out of the inter-crystalline layers, which are replaced by warmer and fresh water from below. Part of the brine remains in the cells between the crystals. The final salinity of sea ice is determined by these salt inclusions, captured during ice growth [13, 17]. The ice, formed in this way, is known as columnar-granular ice; according to the classification of N. V. Cherepanov, it is ice of type B1 … B8.

If the young ice sheet is in a state of free drift or, if there are no constant currents in the direction under the ice sheet, the crystals of columnar ice grow without any preferential direction in the horizontal plane. In this case, the ice sheet is transversally isotropic, i.e. its properties are the same in all horizontal directions.

For conditions of soldered ice, when it is stationary relative to the seabed, and if there are constant sea currents under it, oriented crystal growth is observed. The mechanism of preferential orientation of the C-axes of growing crystals along the flow direction has not yet been sufficiently studied. The degree of ordering of the C-axes usually increases with the increase in the thickness of the ice cover [10–12]. For example, a typical annual soldered ice sheet consists of granular ice with a thickness of about 30 cm, followed by a layer of columnar ice with a more or less random orientation with a thickness of about 50 cm, and finally a layer of columnar ice with a predominant orientation of the C-axes. Often, the lower layers consist of crystals, stretched downwards, the ordering of the C-axes of which fits into a scatter in the direction not exceeding ±5°.

Under unstable conditions of ice formation, a more complex picture of the vertical structure of annual ice is observed (Fig. 2.1). Inclusions of initial ice forms in the form of ice needles and plates, fragments of columnar ice can occur on any horizon. As a result of freezing processes, a natural composite material is created [18–20]. The process of seawater infiltration is also added here, which mixes with the heated areas of the snow cover. Due to the presence of a large amount of snow precipitation, infiltration ice is especially common (see Fig. 2.1) occurs in the Southern Ocean.

Near the shore, the ice is usually smooth and homogeneous at a fairly large distance from it; the monotony of the ice is broken only by extended fractures. This leads to the formation of a homogeneous crystal structure (see Fig. 2.1a).

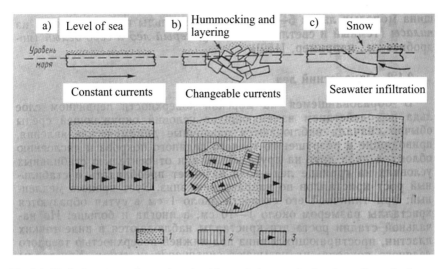

Fig. 2.1 Vertical structure of annual sea ice (diagram). 1—granular ice, 2—undirected columnar ice, 3—oriented columnar ice

The ice cover, farther from the shore, is more affected by variables influence in the direction of winds and currents. The internal structure of the ice, formed as a result of the already mentioned dynamic phenomena is shown in Fig. 2.1b.

Sometimes it is found, that sea ice is not formed from pure seawater, but also includes some sedimentary rocks of various origins, such as sand, gravel, fine clay.

Typical temperature profiles for annual ice are shown in Fig. 2.2.

In winter, the upper surface of the ice sheet (a layer approximately h = 10–20 cm thick) differs by на ± 10 °C from the average air temperature, and the lower layers of ice in contact with water have a stable temperature, equal to its freezing temperature.

Fig. 2.2 Typical temperature profiles in annual Beaufort Sea ice. 1—in winter (February–March), 2—in spring (April–May)

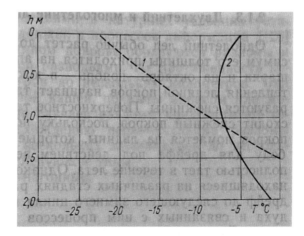

In spring, the air temperature rises, which leads to a corresponding change in the temperature profile of the ice [21]. The gas content (usually it is air) in the volume of ice on average ranges from 0.5 to 5.0%.

Gas, salt, and brine inclusions are the main reason for the different density of sea ice. However, the variability of the annual ice density is small and is in the range of 915–920 kg/m^3 for the middle layers [22]. Only the upper layer with a thickness of about 20 cm has a lower density of 890–920 kg/m^3. The vertical distribution of salinity in the ice thickness is complex, since it depends on the rate of ice build-up.

During the rapid freezing processes of seawater, ice captures a larger amount of brine in the form of closed cells. Sometimes the surface layers of annual ice are saltier (8–12‰) than the underlying ones (5–8‰), which were formed during slower water crystallization processes [20–24]. It has been observed, that as a result of thermodynamic influences, brine cells migrate: when cooling, they fall lower, when warming, they rise up. Their typical geometric shape is vertically elongated closed cavities, which can be combined into brine ducts with a diameter of 1–10 mm.

During hummockage, due to the frontal compression of annual ice floes, hummocks are formed, the surface part of which rises 1–4 m above sea level, and their underwater «height» averages 5–10 m. Hummocks with a draft of several tens of meters are very rarely observed.

2.1.3   Two-Year and Long-Term Ice

Annual ice usually grows to a thickness of 1.0–2.5 m and its maximum thickness occurs in April–May in the northern hemisphere and in October–November in the southern hemisphere. As the seasonal warming, the ice sheet begins to melt, snowflakes form on its surface. Surface melting accelerates as soon as the snow cover disappears, as the albedo decreases. The ice sheet breaks into ice floes, which acquire greater freedom to drift under the influence of wind and currents. Part of the ice floes completely melts during the summer [25]. However, at high latitudes, ice floes at various stages of destruction continue to drift until the next winter cycle of lowering air temperature and the associated processes of ice cover formation. Ice, which has existed for one summer season, is called two-year ice. It is fresher than an annual one, because, in the summer, the brine flows down through drainage channels in the ice structure. The typical salinity of such ice is 1–4‰. As a result of metamorphism processes, ice often turns into a porous granular body.

The thickness of two-year old ice is 2–3 m. According to the WMO classification of sea ice, long-term ice includes those that have existed for at least two summer seasons. Drifting Arctic ice is 5–10 years old. Many researchers believe, that it is quite difficult to distinguish two-year old ice from long-term one. Further thermal and dynamic effects on the two-year old ice cover lead to the fact, that the hummocked areas on its surface become smoothed. In the Central Arctic, the drifting ice sheet almost entirely consists of more or less cohesive long-term ice, dissected by an intense system of cracks, fresh hummocked areas of young ice and a small number of open spaces of the water surface.

The degree of ice destruction during melting is assessed visually on a five-point scale. With a destruction of 0 points, there are no external signs of melting, with a destruction of 5 points, the ice is on the verge of complete destruction. Its structure is honeycomb-like and it is called rotten ice [23, 26].

For long-term ice in the Arctic, the so-called equilibrium thickness is characteristic, equal to 2–6 m, depending on the latitude of the place, and determined, when the additional thicknesses of annual melting and annual freezing are equal. In the structure of the vertical ice core, several «rings» are observed—annual layers 30–50 cm thick in accordance with the age of the ice. The salinity of long-term ice is in the range of 0.5–4.0‰.

2.1.4 Channels and Dilutions, Wormwoods

Numerous observations in full-scale conditions have established, that as a result of movements and deformation processes in a cohesive continuous ice cover, faults and ruptures are formed, which are called dilutions. Their length ranges from several meters to several kilometers. They can be filled with ice porridge, covered with nilas or young ice. In most cases, faults precede hummock formation. By origin, the initial faults in the form of cracks with a width of several millimeters to several meters are divided into the following types [27]:

(1) tidal or run-up;
(2) thermal;
(3) isostatic;
(4) dynamic.

The greatest influence on the formation of the relief of the ice cover is exerted by dynamic cracks, which, as a result of drift and movement of ice, can easily turn into dilutions and channels. There is a narrow dilution with a width of 0–50 m, a small water channel with a width of 50–200 m, an average dilution with a width of 200–500 m and a large dilution with a width of more than 500 m. A channel is any crack or passage through sea ice [28].

The wormwoods are usually referred to as «stable spaces of clear water among stationary ice or on their border. The wormwoods can be filled with ice porridge or covered with initial types of ice, nilas, and young ice. A wormwood, on the one hand, bounded by the coast, is called a coastal wormwood, bounded by stationary ice—a zapripaynaya wormwood. A wormwood that appears in the same place every year is called a stationary wormwood». The formation of wormwoods occurs under the combined influence of currents, tidal phenomena, upwelling, and winds. Most often, wormwoods are observed near the shores and are associated with divergent phenomena in the movement of the ice sheet [25–27].

2.1.5 Ice Drifting Islands and Icebergs

Ice drifting islands are large, on average, flat–parallel formations (pieces) of floating ice, rising above sea level by 5 m or more, which break off from Arctic ice shelves; they have a thickness of 30–50 m and an area of several thousand m^2 to 500 km^2 or more.

Most of the surface of these islands consists of ice, formed from snow and lake ice, which is the result of compaction of snow that has fallen as precipitation, and the re-freezing of snowfields. However, the lower part of the island's thickness is typical sea ice or brackish ice with unusually large crystal sizes, which apparently indicates very slow processes of ice freezing from below at the ice–water boundary. It is believed that shelf glaciers (they can also be attributed in some cases to soldered long-term ice sheets, several tens of meters thick), producing ice islands are in an equilibrium state between the ocean and the atmosphere, having a balance of melting and accumulation processes [29].

The existence of ice islands in the Arctic has been known for a long time. At one time, scientific research drifting stations were organized on some of them. So, on the ice island of J. Fletcher 70 m thick, discovered in 1950, was planted by the American T-3 station, which was able to be used for its intended purpose for more than 20 years, since it fell into the anticyclonic Beaufort drift cycle. Soviet drifting stations were also organized on ice islands, several tens of meters thick [30].

There are no clear statistics on the number and size of ice islands in the Arctic basin. However, according to some data, in 1972, about 433 ice islands were noted along the coast of the Beaufort Sea. Of these, 117 had transverse dimensions, exceeding 80 m, and one had a diameter of over 1600 m.

Icebergs are distributed over an area of about 20% of the World's oceans, and more than 90% of their total number are located in the southern hemisphere. According to the current International nomenclature of sea ice, an iceberg is called «a massive piece of ice of various shapes that has broken off from a glacier, protruding more than 5 m above sea level, which may be afloat or stranded. In northern latitudes, the transverse size of a large iceberg is about 200 m, and the elevation above sea level is 25.5 m, the thickness of the underwater part, which accounts for 90% of the total volume of ice, reaches 225 m. Since most of the iceberg is under water, its movement is almost entirely determined by sea currents. Thus, Greenland icebergs are carried into the Atlantic Ocean by the East Greenland current» [31].

A large number of large icebergs are found in the Southern Ocean and adjacent waters. The area of the Antarctic ice sheet is approximately 14.0 million km^2, and the total volume of its ice is 23–30 million km^3. Ice formations, reaching several kilometers across and up to 500 m thick, break off from the outlet glaciers in the peripheral zone. The lifetime of the bulk of icebergs, which are relatively small in size, is estimated at 35 years. The direction of movement of Antarctic icebergs coincides with the drift of pack ice in these areas. However, they can cross the band of the famous «roaring forties» and penetrate far north into the Pacific and especially into the Atlantic Oceans. The melting of such large icebergs can last up to 10 years.

Table 2.2 Classification by iceberg size

Name	Elevation above sea level (m)	Length (m)	Approximate weight (ton)
Growler	<1.5	<5	100
Chip	1.5–5.0	5–15	1,000
Little	5–15	15–60	100,000
Average	15–50	60–120	2,000,000
Big	50–100	120–220	10,000,000
Very big	>100	>220	>10,000,000

Above sea level, we see only the tip of the iceberg. Its bulk is under water. The ratio of height above sea level to its sediment strongly depends on the shape and density of the iceberg, the average value of which varies in the range of 850–910 kg/m^3. The surface part of Antarctic icebergs consists mainly of lighter snow and firn, their average density does not exceed 800 kg/m^3. Greenland icebergs, as a rule, have a small layer of snow or none at all, since the surface is usually subjected to intense melting and refreezing before the iceberg breaks off from the glacier into the sea [31–33]. The shape of the iceberg also affects the ratio of average height to draft. As a rule, this ratio is in the range from 1:4 to 1:6. Classification of icebergs by size with indication of their approximate mass is given in Table 2.2.

2.2 Structure of the Sea Ice Cover and Dynamic Processes, Occurring in It

The sea ice cover can be stationary and mobile. This is of fundamental importance, when determining the location for the construction of ice airfields and, especially, when placing radio equipment for flight support on them. In the first case, it is a solder—a continuous cover (with possible continuity violations, for example, in the form of cracks and hummocks of thermal origin), connected to the shore. In the second case, it is drifting ice that is not connected to the shore; the reason for their movement is usually wind and currents. By area, they occupy most of the freezing sea areas [25, 29, 33].

The structure of ice, floating on the sea surface, can be considered as an ensemble of individual ice floes (fragments of fields, fields of frost) with some statistical values of the corresponding geometric parameters. Moreover, the manifestation of structural anisotropy or isotropy of drifting sea ice depends on the scale of the studied areas, i.e. the scale of averaging. Small-scale areas of the ice cover and their corresponding movements, commensurate with the size of individual ice floes, are, as a rule, significantly anisotropic [34]. In large-scale phenomena with characteristic dimensions of several hundred kilometers, the ice sheet can practically be considered as an isotropic medium.

2.2.1 Forms of Spatial Structure and Geometric Characteristics of the Ice Cover

The following stable forms of spatial structure are characteristic of the ice cover: polygonal, grid, spotted and branched (dendritic). In winter, in the Arctic basin, the main elements of the cover are polygonal ice blocks, violations of the continuity of sea ice (faults, ridges of fresh hummocks, cracks, channels) form a polygonal system.

At the transition stage from winter to summer, the spatial structure of the ice sheet becomes reticulated. During this period, numerous cracks, dividing the ice sheet into small ice formations, appear. The fragmentation of the ice cover increases even more in summer, the area of «clean water» increases significantly, compared to the area of the dilutions and primary forms in winter. The length of the ice fields in summer rarely exceeds 1 km, and they are partly rounded in shape. Among the rare and sparse ice floes of smaller sizes, there are bands and spots of more cohesive and larger ice floes [31]. This form of the spatial structure of the sea ice cover is called spotted, it is characterized by vortex formations in the form of so-called ice spirals. And, finally, in autumn, the dendrite spatial structure of the ice cover is most often found, which got its name from the nature of branching cracks in the young ice of freshly frozen water spaces.

The main geometric characteristics of the ice cover include both linear (ice thickness, transverse size and perimeter of the ice floe, crack length, etc.) and non-linear (the area of ice floes and dilutions, the volume of ice, for example, in hummocks, the cohesion of ice). By size, drifting ice is divided into two main groups: ice fields and broken ice. These forms are formed as a result of the destruction of the solder into fragments and their subsequent partial crushing, the formation of young ice, the formation of frost by freezing individual ice floes, etc. The largest formations of sea ice are ice fields with transverse dimensions: more than 10 km (giant), 2–10 km (extensive) and 0.5–2 km (large). The chips of fields include ice floes with characteristic dimensions of 100–500 m. Smaller ice floes form broken ice, which in turn is divided into coarse-beaten (20–100 m), fine-beaten (less than 20 m), and grated ice (less than 2 m). Ice islands and icebergs are allocated to a separate class.

Ice floes, in their outlines and from a geometric point of view, can generally be considered as convex flat simply connected figures of arbitrary shape and size, covering the surface of the sea without intersections [35].

2.2.2 Cohesion and Fragmentation

The main integral characteristic of drifting ice is its cohesion. It is determined by the ratio of the area of ice floes on the observed section of the sea to the area of this section. Usually, cohesion is measured on a ten-point scale. Zero points correspond to clear water, and ten points correspond to solid ice. The cohesion of the ice is determined instrumentally and visually. When processing aerial photography data, counts of n segments length l_i with varying degrees of ice cover are made along the entire scanning line of long L. In this case, the assessment of cohesion will be [22]

$$N_c = \frac{1}{L} \sum_{i=1}^{n} l_i. \qquad (2.1)$$

It can be shown that cohesion is related to the drift velocity vector of ice floes $\overline{V}$ (under the condition of small movements, occurring over limited time intervals δt) by the relation [23]

$$\frac{1}{N} \frac{\delta N}{\delta t} = -\mathrm{div}\,\overline{V}. \qquad (2.2)$$

The characteristic of fragmentation of the ice cover is determined by the fragmentation score (by the number of discontinuities in the observation area with a length of at least 50 km in winter or by the number of ice fields, chips, and broken ice with a smaller averaging scale of about 10 km in summer) and the function of distribution of ice floes by size.

The fragmentation of the ice cover of the studied water area can also be described by the frequency of recurrence $P_i = P(S_i)$ of ice floes of a certain area S_i. At the same time, seven to nine grades are usually set for the size of ice floes. In practice, the distribution function $F(S_i)$ is often used. It is the ratio of the area, made up of ice floes of size S_i to the total area of the S_{av}, covered with ice. This probabilistic function with a discrete distribution of ice by class is written as [26]

$$F(S_i) = \frac{P_i S_i}{\sum_{i=1}^{n} P_i S_i} = \frac{P_i S_i}{S_{av}} \qquad (2.3)$$

2.2.3 Distribution of Ice Thickness

It is established, that the thickness of the ice has not only large-scale, but also local variability. The existing variability at a distance of several meters is due to thermodynamic processes. Moreover, with increasing age of ice, the range of thickness changes increases.

The distribution of thicknesses on the considered area S is conveniently described by the function $g(h)$, according to the expression [17]:

$$\int_{h_1}^{h_2} g(h)dh = \frac{1}{S} A(h_1, h_2), \qquad (2.4)$$

where A—area, covered with ice thickness $h_1 \leq h < h_2$.

The reason for the changes in $g(h)$ are mainly two processes: thermodynamic and mechanical. The first of them causes a change in the thickness of the ice from above and below and establishes an equilibrium thickness. The second leads to the formation of spaces of water, free from ice, and to the accumulation of hummocks. The thickness distribution at any given time shows the combined effect of these two processes. Currently, it is not difficult to find experimental data in the scientific literature, obtained by remote methods, characterizing the variability in space and time of the thickness of the sea ice cover. For example, from the results of processing aerial photographs of the Central Arctic Basin, it follows, that in summer the most likely amount of long-term ice here is about 75%, annual—10%, young ice and clean water 5–7%. According to sonar profiling, in winter the probability of meeting young ice is only 2%. An idea of the spatial heterogeneity of ice thicknesses of different ages is given by autocorrelation functions, the analysis of which leads to the conclusion, that annual ice and winter young ice are the most even (correlation radius of about 70 m). The correlation radius for long-term ice is approximately 30 m [36].

Figure 2.3 shows the functions of ice thickness distribution that characterize the two-year period of development of the variability of the thickness of the Arctic sea ice cover.

2.2.4 Hummockiness

Compressive and tangential forces, arising during the contact of ice floes with each other, lead to ice breaking and the formation of hummocks. There are wind hummock, tidal hummock and thermal hummock. The hummockiness of the ice cover is practically determined by a five-point scale that evaluates the ratio of the area, covered by hummocks, to the area, occupied by ice in the observer's field of view [37]. The compression, causing hummockiness, is balanced by the floatation forces of the formed fragments, the elastic inertia forces and the forces, developing in the ice plate. The relation between these groups of forces, as well as the thickness and strength of the ice cover determine the shape and size of the hummocks. The analysis of stereo photographs from airplanes gave statistical data on the size, structure and geometric shape of the ridges of hummocks. As their parameters, the average values and variances of height, the ratio of the width of the base to height, the ratio of underwater part to height, the distance between the ridges of hummocks were used. The parameter values for ice of different ages are given in Table 2.3.

It is established, that the average height of the ridges of hummocks, formed on annual ice is 1.6–1.9 m, and on long-term ice is 2.0–2.5 m. The height of the hummocks is related to the geographical location of the area, but the average values will be close to the specified limits. The most important morphological characteristic of hummock ridges is their underwater part. Based on the results of direct measurements by scuba divers, and in some cases by stereoscopic shooting, using special

Fig. 2.3 The functions of
ice thickness distribution

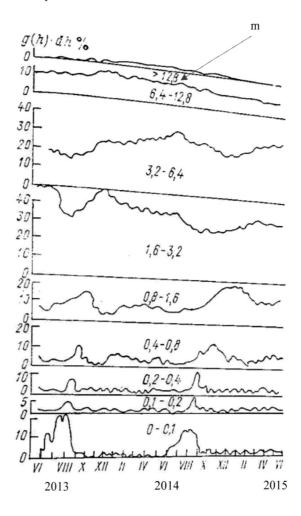

cameras, lowered from the ice surface, it was found, that the average ratio of underwater part to above-water height is in the range of 4.8–5.5. Such an attitude value
was also obtained by American specialists. Moreover, the average value of the ratio
of underwater part to above-water height is preserved for almost the entire range of
heights of the observed hummocks in ice of various ages. Based on the presented
results, a model of the shape of the ridges of hummocks was proposed, according to
which the shape of their cross-section is practically an isosceles trapezoid. Its lower
base is equal to five, the upper one is the same height of the ridge of hummocks.
The underwater part has a draft, equal to 5 heights and a cross-sectional shape in the
form of an inverted isosceles trapezoid with an upper base, equal to seven, and the
lower one is one height of the hummock. The slope angle is 26.5°. Sometimes, the
triangular shape of the cross sections of the above-water and underwater parts of the
ridge was used in model calculations.

Table 2.3 The average values of some parameters of ridges of hummocks and the standard deviations δ from these values

The prevailing age of ice in the area	Height of hummocks (m)		The ratio of the width of the base to the height		The ratio of underwater part to height		The distance between the ridges of hummocks (m)	
	H	δ	b/h	δ	z/h	δ	R	δ
Thin annual ice	1.9	0.7	5.6	2.6	–	–	–	–
Long-term ice	2.5	1.1	3.6	2.1	–	–	–	–
	2.2	0.2	5.4	–	4.8	1.8	–	–
	2.0	1.0	5.1	–	–	–	340	304
Thick annual ice	1.6	0.8	5.0	0.6	5.0	0.8	93	72
	1.9	0.7	2.7	0.6	5.5	1.1	89	67
Long-term ice	2.2	–	1.7	–	5.0	–	111	–

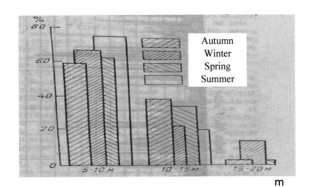

Fig. 2.4 Seasonal distribution of average values of ice underwater part in the central part of the Arctic basin, based on the results of processing of echoledograms

Most of the data on the hummockiness of Arctic ice is obtained as a result of studies of linear profiles or routes, carried out, using submarines and laser profilometers, installed on aircraft. The hummockiness of the ice sheet in a one-dimensional sense can be fully characterized by two parameters: N_T—the average number of hummocks per unit length and h—the average height (or draft, if its underwater part is described). Figure 2.4 and Table 2.4 show experimental data, characterizing the morphological features of drifting ice.

2.2.5   Characteristics of Ice Drift and Compression

When analyzing the ice flow pattern in the Arctic Ocean, two features can be noted: a transpolar drift flow, spreading from the East Siberian Sea through the North Pole to the northeast of Greenland, and an anticyclonic circulation (clockwise) in the area

Table 2.4 The number of channels and dilutions at a distance of 30 miles in various areas of the Arctic basin

Region	January–May	June–July	August–September	November–December
Chukchi Sea	13	11	17	15
Beaufort Sea	8	13	18	12
Canadian Coastal	11	10	12	9
Arctic Ocean				
Eastern part	12	13	17	9
Western part	8	13	14	10
Greenland Sea	10	13	15	–
East Siberian Sea	9	–	16	12

between Alaska, Canada and the North Pole—the so-called Beaufort spiral. The cause of the transpolar drift flow is the cold surface waters of the Siberian shelf. Here, ice arrays are formed due to the split of the coastline or near islands, that make it difficult to drift freely (for example, an ice array is known in a northerly direction off the Siberian coast near the island of Ayon). As a result of northward drift, areas of clear water sometimes open up near the Siberian coast even in winter, on which thin ice forms—the wormwoods with an area of up to several hundred km^2. The drift of the Norwegian vessel «Fram» and the Soviet research stations showed, that the distance from Chukotka to the Greenland Sea ice passes in about 3–5 years [38]. As a result, large masses of polar ice are carried into the North Atlantic mainly through the Danish Strait between Iceland and Greenland. This diagram also shows, that the bulk of the ice gradually deviates from the shores of Eurasia, while some of the ice passes the Laptev Sea, the Kara and Barents Seas, in which the movement of ice follows a local closed trajectory. The nature of ice movement off the coast of North America is completely different. The oldest and heaviest ice in the Arctic is located in this area; ice fields in the center of rotation can exist here for more than 20 years. These old ice fields contain a large number of smoothed old hummocks. Ice, moving along the outer side of the Beaufort spiral, makes a complete revolution in 10 years, and near the center of the spiral—in 3–4 years. Ice fields from the outer edge of the circulation can be involved in the transarctic drift from east to west and carried out into the Atlantic [37]. Numerous observations of research stations have shown, that the direction of their path on the ocean surface is extremely broken. The average annual drift speed ranges from 0.4 to 4.8 km/day, and the actual speed (including loops and other deviations from the general drift) reaches 2.2–7.4 km/day. In short periods of time, the average monthly speed reaches 10.7 km/day. And in the southern part of the cycle, a high speed of ice drift was recorded, equal to 7 km/h.

The most difficult problem of ice cover dynamics is the description of the interaction of ice floes with each other. This is due to the fact that various physical processes occur during the interaction in the ice. These internal processes have been studied extremely poorly. In nature, ice rarely drifts in a single stream. Due to the dissimilarity of forms, the unevenness of wind and currents, the impact of the shores, ice floes move at different speeds. In rare and sparse ice, debris during drift bends around large fields, their collision and rotation occur. Cracks, breaks of ice fields, dilutions are formed in the solid ice cover.

The resulting anomaly in the movement of one of the ice floes, caused, for example, by wind or current, is transmitted to others as a result of hydrodynamic interaction through collision (pulse exchange), pressure and edge friction (voltage interaction). Thus, ice floes continuously exchange pulse, and the features of their interaction depend on the cohesion of the ice cover. The possibility of pulse transfer makes the ice environment is connected in the sense, that the drift in one of the areas of space is determined not only by external forces, acting at a given point, but also by the dynamic state of the ice cover in adjacent areas.

The hydrodynamic theory of ice sheet drift is based on the equation of the balance of forces, acting on a unit of ice surface [39]:

$$\rho H \frac{\partial \overline{V}}{\partial t} = \overline{\tau_a} + \overline{\tau_\omega} + \overline{F_C} + \overline{F_g} + \overline{F_r} \tag{2.5}$$

where ρ—ice density; H—ice thickness; $\overline{V}$—drift velocity vector speed; τ_a—tangential wind pressure on the air-ice partition surface; τ_ω—friction of ice on water; $\overline{F_C}$—the power of Cornolis; $\overline{F_g}$—gravity projection on the sea surface; $\overline{F_r}$—internal resistance, caused by the interaction of ice fields.

And, if satisfactory methods of estimation are found for the first four terms, then the last term, for the above reasons, causes significant difficulties in solving this equation. For example, in one of the models (Doronin, Egorov, Laichtman, Campbell), according to which ice drift is represented as the movement of a layer of a very viscous liquid, located between less viscous liquids-air and water, the force of interaction is identified with the viscous forces, acting in liquids, and is expressed by the equation

$$\overline{F_r} = \rho H k_x \nabla^2 \overline{V} + k_x \nabla^2 \overline{V}, \tag{2.6}$$

where $\nabla^2 = \frac{\partial^2}{\partial x^2} + \frac{\partial^2}{\partial y^2}$—two-dimensional Laplace operator, k_x—the kinematic coefficient of horizontal turbulent viscosity, which, by analogy with the diffusion coefficient of smoke in the atmosphere, can be written as $k_x = \Delta l^2 / \Delta t$, where Δl—distance change over time Δt.

He also proposed to use the following four equations to describe the spatial unevenness of ice drift:

$$\begin{aligned}
\operatorname{div} \overline{V} &= \partial v/\partial y + \partial u/\partial x \\
\operatorname{rot} \overline{V} &= \partial v/\partial x - \partial u/\partial y \\
\operatorname{def_1} \overline{V} &= \partial v/\partial x + \partial u/\partial y \\
\operatorname{def_2} \overline{V} &= \partial v/\partial y - \partial u/\partial x
\end{aligned} \tag{2.7}$$

where $\overline{V}$—drift velocity vector, v and u—projections $\overline{V}$ on the x and y axes, respectively.

The first equation (divergence) characterizes the intensity of expansion (compression) of the ice sheet at a given point, the second (rotation)—the speed of its rotation. Both of them are invariant with respect to the choice of the direction of the coordinate axes. The last two equations characterize the rate of change in the shape of the ice sheet element, i.e. deformation. In this case, the third equation expresses angular deformation, and the fourth—the unevenness of linear deformation along two perpendicular axes. Expression [35]

$$\operatorname{def}\overline{V} = \sqrt{\left(\operatorname{def}_1\overline{V}\right)^2 + \left(\operatorname{def}_2\overline{V}\right)^2} \qquad (2.8)$$

is also invariant with respect to the choice of coordinate axes and therefore objectively shows the change in the shape of the ice cover element. If we had the possibility of sufficiently accurate experimental measurement of the spatial deformation characteristics of ice and their changes over time, then the problem of studying the dynamics of sea ice would be facilitated. The task of studying the medium-scale characteristics of the dynamics of ice sheets is solved, in particular, by a method, based on the principles of radio hydroacoustics.

Considering linearized models of ice compressions, which are represented by forced periodic unsteady processes, it can be shown, that the physical mechanism of compression propagation in cohesive ice depends on their scale. Thus, with horizontal sizes of compression regions of about 100 km or less, the compression propagation is determined by quasi-elastic forces of interaction between ice floes [39–41]. Inertial forces play a significant role in the propagation of large-scale disturbances. The presence of constantly acting dissipative factors leads to the attenuation of ice movements. For small-scale movements, this occurs as a result of lateral pulse exchange. Movements of a larger scale are attenuated mainly due to the transfer of energy to the subglacial layer of water.

Here are some calculated characteristics, summarized in Table 2.5 to illustrate the phenomenon under consideration.

Currently, along with the three-point scale of visual assessment of sea ice compression, used in practice, the search for ways of instrumental assessment of this phenomenon is continuing. One of the directions is the development of a method, based on the direct measurement of deformation and stress in the ice cover with the transmission of data over the radio channel to the information receiving point for further processing.

Table 2.5 Characteristics of ice compressions in the sea ice cover with an external tangential force, equal to approximately 0.01 Pa

Compression propagation speed (m/s)	Characteristic size of the compressed area (km)	Compression force (H/m)	Note
1	10^1	10^3	Elastic internal forces can be neglected
1	10^2	10^2	Elastic internal forces can be neglected
10	10^1	10^5	The value is close to the limit, recorded when ships are compressed in ice
10	10^2	10^4	The value is close to the limit, recorded when ships are compressed in ice

References

1. Kozlov AI, Sergeev VG (1998) Propagation of radio waves along natural routes. Moscow
2. Kozlov AI, Maslov VY (2004) Differential properties of the scattering matrix. Sci Bull Mosc State Tech Univ Civ Aviat 79:43–46
3. Shatrakov YG (2018) Development of domestic radar aircraft systems. Publishing House Stolichnaya Encyclopedia, Moscow, p 400
4. Drachev AN, Farafonov VG, Balashov VM (2014) Methods of control of complex profile surfaces. Radio Electron Issues 1(1):91–99
5. Antsev GV, Bondarenko AV, Golovachev MV, Kochetov AV, Lukashov KG, Mironov OS, Panfilov PS, Parusov VA, Raisky VL, Sarychev VA (2017) Radiophysical support of ultrashort pulse radar systems. In: Problems of remote sensing, propagation and diffraction of radio waves. Lecture notes. Scientific Council of the Russian Academy of Sciences on Radio Wave Propagation, Murom Institute (Branch), Vladimir State University, Named After Alexander Grigoryevich and Nikolai Grigoryevich Stoletov, pp 5–21
6. Eliseev BP, Kozlov AI, Romancheva NI, Shatrakov YG, Zatuchny DA, Zavalishin OI (2020) Probabilistic-statistical approaches to the prediction of aircraft navigation systems condition. Springer Aerospace Technology, p 200
7. Kaplun VA (2004) Radio-transparent antenna fairings. Antennas 8–9(87–88):109–116
8. Kravchenko NI, Bakumov VN (1999) The limiting error of measuring the regular Doppler shift of the frequency of meteorological signals. News Univ: Radioelectron 4:3–10
9. Kravchenko NI, Lenchuk DV (2001) The ultimate accuracy of measuring the Doppler shift of the weather signal frequency, when using a bundle of coherent signals. News Univ: Radioelectron 7:68–80
10. Krasyuk NP, Koblov VL, Krasyuk VN (1988) The influence of the troposphere and the underlying surface on the operation of radar systems. Radio and Communications, Moscow, p 216
11. Krasyuk NP, Rosenberg VI (1970) Ship radar and meteorology. Shipbuilding, Leningrad, p 325
12. Cook C, Bernfeld M (1971) Radar signals (translated from English). Soviet Radio, Moscow, p 568

13. Kulikov EI (1964) Limiting accuracy of measuring the central frequency of a narrow-band normal random process against a background of white noise. Radio Eng Electron 10:1740–1744
14. Levin BR (1989) Theoretical foundations of statistical radio engineering. Radio and Communications, Moscow, p 656
15. Marple SL Jr (1990) Digital spectral analysis and its applications (translated from English). Mir, Moscow, p 584
16. Melnik YA, Stogov GV (1973) Fundamentals of radio engineering and radio engineering devices. Soviet Radio, Moscow, p 368
17. Melnikov VM (1997) Meteorological informativeness of Doppler radars. In: Proceedings of the All-Russian symposium «radar studies of natural environments», issue 1. Military Engineering Space Academy, Named After A. F. Mozhaisky, St. Petersburg, pp 165–172
18. Melnikov VM (1993) Information processing in Doppler weather radars. Foreign Radio Electron 4:35–42
19. Minkovich BM, Yakovlev VP (1969) Theory of antenna synthesis. Soviet Radio, Moscow, p 296
20. Mironov MA (2001) Estimation of the parameters of the autoregression model and the moving average, based on experimental data. Radio Eng 10:8–12
21. Vereshchagin AV, Zatuchny DA, Sinitsyn VA, Sinitsyn EA, Shatrakov YG (2020) Signal processing of airborne radar stations plane flight control in difficult meteoconditions. Springer Aerospace Technology, p 218
22. Ostrovityanov RV, Basalov FA (1982) Statistical theory of radar of extended targets. Radio and Communications, Moscow, p 232
23. (1983) Problems of creation and application of mathematical models in aviation. Series «Questions of cybernetics». Nauka, Moscow
24. Melnik YA (ed) (1980) Radar methods of Earth research. Soviet Radio, Moscow, p 264
25. Kondratenkov GS, Potekhin VA, Reutov AP, Feoktistov YA (1983) In: Kondratenkov GS (ed) Radar stations of the Earth survey. Radio and Communications, Moscow, p 272
26. Sarychev VA, Antsev GV (1992) Modes of operation of multifunctional airborne radar systems for civil purposes. Radioelectron Commun 4:3–8
27. Guterman IG (ed) (1968) Handbook of climatic characteristics of the free atmosphere for individual stations of the Northern hemisphere. Moscow
28. Tikhonov VI, Kulman NK (1975) Nonlinear filtering and quasi-coherent signal reception. Soviet Radio, Moscow, p 704
29. Uskov V (1987) Wind shear and its effect on landing. Civ Aviat 12:27–29
30. Feldman YI, Mandurovsky IA (1988) Theory of fluctuations of location signals, reflected by distributed targets. Radio and Communications, Moscow, p 272
31. Yurkov NK, Bukharov AY, Zatuchny DA (2021) Signal polarization selection for aircraft radar control—models and methods. Springer Aerospace Technology, p 140
32. Veiber YE, Skachkov VA, Smirnov NK (1987) Features of measuring the velocities of relative movements of atmospheric formations by radar methods. Academy of Sciences of the USSR, Moscow, p 13
33. Bean BR, Dutton ED (1971) Radiometeorology. Hydrometeorological Publishing House, Leningrad, p 362
34. Brylev GB, Gashina SB, Nizdoiminoga GL (1986) Radar characteristics of clouds and precipitation. Hydrometeorological Publishing House, Leningrad, p 231
35. Van Tris G (1977) Theory of detection, evaluation and modulation, vol 3 (translated from English). Soviet Radio, Moscow, p 664
36. Vereshchagin AV, Mikhailutsa KT, Chernyshov EE (2002) Features of detection and evaluation of characteristics of turbulent weather formations by onboard Doppler weather radars: report at the XVIII All-Russian symposium «radar research of natural environments» (18–20.04.2000). In: Proceedings of the XVI-XIX All-Russian symposium «radar research of natural environments», issue 2. St. Petersburg, pp 240–249
37. Vityazev VV (1993) Digital frequency selection of signals. Radio and Communications, Moscow, p 239

38. Zatuchny DA (2017) Analysis of the features of wave reflection, when transmitting data from an aircraft in urban conditions. Informatiz Commun 2:7–9
39. Zatuchny DA, Kozlov AI, Trushin AV (2018) Distinguishing observation objects, located within the irradiated surface area. Informatiz Commun 5:12–21
40. Alekseev VG (2000) On nonparametric estimates of spectral density. Radio Eng Electron 45(2):185–190
41. Andreev VG, Koshelev VI, Loginov SN (2002) Algorithms and means of spectral analysis of signals with a large dynamic range. Quest Radio Electron: Ser Radar Equip (1–2):77–89

Chapter 3
Physical Properties of Ice, Used to Solve Problems of Civil Aviation

3.1 Atomic-Molecular Structure of Ice and Phase Diagram of Sea Ice

Of the ten known forms of ice and its amphoric states (obtained, among other things, in laboratory conditions at high pressure, reaching several gigapascals), the most studied is the so-called hexagonal ice, or ice *Ih*. This is ordinary ice with a hexagonal packing of atoms, which is formed from freshwater at a temperature of about 0 °C and atmospheric pressure (in contrast to the cubic modification of ice *Ic*, which is in a metastable state with respect to ice *Ih* in the temperature range—140 to −120 °C and also at atmospheric pressure) [1].

Ice *Ih* is the main component of the sea ice cover as a multiphase system, formed from salty seawater and containing brine and gas inclusions in addition to pure ice crystals [2].

3.1.1 Crystallographic Structure of Ice

The crystal structure of ice is caused by tetrahedral coordination of oxygen atoms. Moreover, an important feature of its hexagonal-symmetric molecular structure is that all molecules are concentrated in closely packed parallel planes, called basic planes. The direction perpendicular to the basic plane is called the C-axis of the crystal. Ice *Ih* has an open structure, so its density is less than the density of water.

The main building block of the «crystallographic» structure of ice is an ensemble of atoms, occupying a region of space, known as an elementary cell, from which ice is formed. In an elementary cell, the total number of oxygen atoms (and hence the number of H_2O molecules) is four. The tetrahedral angle is 109°28′, the distance between neighboring molecules in ice *Ih* at 0 °C is 2.76 Å. The elements of the ice elementary cell fit into a spatial lattice in the form of a parallelepiped, the eight vertices of which are called lattice nodes [3–6].

© The Author(s), under exclusive license to Springer Nature Singapore Pte Ltd. 2023
A. I. Kozlov et al., *Ice Structures for Airfield Construction*, Springer Aerospace
Technology, https://doi.org/10.1007/978-981-19-6211-0_3

The dimensions of the lattice are determined by the length of the rhombohedral base a_0 and the height of the lattice c_0. The parameters of the elementary cell vary depending on the temperature of the ice, but the ratio $c_0/a_0 \approx 1.63$. The crystallographic features of the symmetry of the Ih lattice are as follows: it belongs to a primitive, or simple, type of lattices; its main axis of symmetry (C-axis) is a hexagonal helical; it has two rows of symmetrical planes: mirror and sliding planes; the molecular structure of ice Ih is centrally symmetric, i.e. in the ice lattice, we can always specify a point, relative to which there are molecules, located at the same distance in opposite directions from it. So, the structure of ice is determined by the structure of water molecules.

It is known that three rows of H_2O molecule form are an isosceles triangle by covalent bonding with oxygen nuclei at the apex and two protons at the baseline. The nuclei of the water molecule are surrounded by an electron cloud, consisting of 10 electrons, arranged unevenly inside the sphere. Two of them are in the first orbit, in close proximity to the oxygen nucleus, and do not take a noticeable part in the formation of a bond between oxygen and hydrogen, while eight other electrons rotate in pairs in four eccentric orbits, directed tetrahedrally from the oxygen nucleus. Protons create two positive centers. The charge of eight electrons completely compensates for the charge of the oxygen nucleus, but the electrons, rotating in the direction of two orbits, that do not contain protons, form two negative centers, i.e. the uninhabited electrons form, as it were, two «hands», stretched from the oxygen nucleus to the vertices of an imaginary tetrahedron, in which the H_2O molecule is inscribed [2, 3, 5].

«In the case, when water molecules are not isolated, but collected in a small volume close enough to each other, as in a liquid, these negatively charged «hands» contribute to the attraction of molecules among themselves. Each negatively charged «hand» attracts a hydrogen nucleus from a neighboring water molecule, and hydrogen atoms, therefore, contribute to the connection of molecules through the so-called hydrogen bond. Molecules tend to form groups of molecules (clusters). In a liquid, usually the thermal motion of molecules is sufficient to break the bonds of molecules: clusters are continuously crushed and transformed. However, when the water cools to the freezing point, the thermal motion becomes so weak, that the molecules form large stable clusters: ice crystals» [7].

The forces, forming the hydrogen bond, are essentially electrostatic in nature. A simple explanation of the bonds and tetrahedral arrangement of Ih ice molecules can be obtained, using the H_2O orbital model, consisting of point charges. According to this model, in a tetrahedral system, formed by an oxygen nucleus O, surrounded by four molecules, the attractive force, acting from the side of unbound electrons on the proton H forms a hydrogen bond between the molecules. This attractive force increases the O-H distance from 0.9 A (in the water vapor phase) to 1.011 A (in the Ih ice crystal lattice). Under the action of attractive forces and unbound electrons, a proton can occupy any of the two positions of the potential minimum at the corresponding points of the orbital model [8]. Taking into account the known interaction forces (electrostatic, dispersion, delocalization), it was possible to calculate the total energy of the hydrogen bond. It was 23.9 kJ/(kg · K). According to other data, this energy is equal to 19.6–34.3 kJ/(kg · K). These estimates correspond to the intermediate value of the dissociation energy of typical covalent bonds.

3.1.2 Dynamics and Disruption of the Structure of the Crystal Lattice of Ice

In the lattice of an ice crystal, as in a rigid fixed structure, natural oscillations can occur, called normal modes, if we understand by this such oscillations, in which all nuclei oscillate with the same frequency and in the same phase. If some disturbance is applied to the lattice (for example, light radiation, an electron beam, etc.), then the oscillatory process in it will become more complicated. The crystal always has its own thermal vibrations, which can also be represented in the form of waves—acoustic and optical photons. However, the random arrangement of hydrogen atoms in their molecule and the equally probable presence in a neighboring molecule generally make it possible to describe vibrational excitations in the ice lattice, using photons [9–11]. To isolate and study the properties, introduced by the presence of a lattice, it is necessary to conduct a detailed comparison of free H_2O molecules and molecules, that make up the ice crystal. A free water molecule has nine degrees of freedom: three translational, three rotational, and three vibrational (Fig. 3.1).

Since the motion of the H nuclei during the oscillations of υ_1 and υ_3 occurs along the direction of the O-H bonds, these modes are called O-H bond stretching oscillations. With oscillations of type υ_2, the nuclei of H move perpendicular to O-H, hence υ_2 is called the deformation oscillation of the H-O-H bond, or the bending oscillation of the H-O-H bond. The values of frequencies υ_1, υ_2, and υ_3 are obtained by studying infrared spectra.

Currently, there is a theory of vibrations of molecular crystals, that describes in sufficient detail the vibrations of bound water molecules through their potential functions.

Ice does not have an ideal structure—there are defects in its lattice, the presence of which mainly determines the properties of solids, that we observe in practice. Inclusion in the form of an extraneous molecule of ice or other substance, placed in the internode, moves to neighboring cavities by jumps with a frequency of 10^6 s^{-1} on average at a distance of up to eight molecular radii until it takes place in the lattice. It is possible to replace water molecules with other molecules, capable of forming a hydrogen bond. However, the main interest in considering ice as a molecular crystal is the possibility of the formation of orientation defects. In 1951, Bjerrum proposed to call L-defects a situation, when there are no protons on the hydrogen bond line, D-defects, when there are two protons on the hydrogen bond line. The existence of L- and D-defects allows the molecules to rotate, and the rotation speed reaches 628×10^3 рад/s. This possibility causes *the polarization of the ice*. The conductivity of ice is ionic in nature, it is convenient to represent ions in the form of defects, associated with the Bjerrum defect, i.e. to consider hydronium H_3O^+ and gridroxyl

Fig. 3.1 Normal modes of
vibration of a water molecule

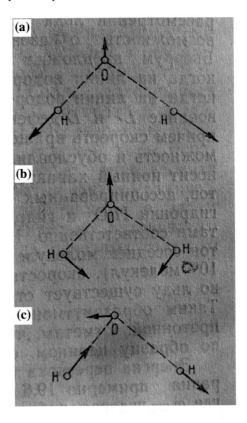

OH⁻ as molecules with D- and L-defects, respectively. Hydronium can then transfer the excess defect to the average molecule, etc. Although there are few such ions (one per 10^{12} molecules), the speed of their movement is very high, since there is a structure in the ice, that fixes the position of neighbors. Thus, ionic conductivity is actually protonic. Note, that in this case, no particles move along the sample entirely, there are only local displacements of defects [7–9, 12].

The energy of the proton jump from the far to the near position is approximately 19.6 kJ / (kg · K), the jump time is equal to π/ω, ω is the angular frequency of the proton. The electric field of the L-defect has a compensating effect (its effective charge is 0.6e) and allows multiple collisions of protons and corresponding mixing of the L-defect in this region. It is necessary to take into account the average number of molecules, involved in the processes of rotation and collision. It depends on the electric field, the activation energy, and the dissipation of rotational energy in the lattice. The conductivity, mobility, and diffusion of the L-defect are proportional to the square of the jump length of the defect in the thermally activated initial motion process, increasing linearly with an increase in the number of molecules. Impurities, placed in the ice lattice, change the electrical and mechanical properties of ice. For example, the addition of HF produces the effect of significant softening of ice, and NH_3 causes its slight hardening.

3.1.3 Sea Ice Phase Diagram

Since the sea ice cover as an object of remote studies is a multicomponent medium, containing, in addition to pure ice crystals, solid, liquid, and gaseous inclusions, that significantly determine the range of variability of its physical properties, it seems necessary to consider the main features of the phase composition of sea ice, or its phase diagram, which in this case means the same [13].

It is known that seawater, unlike freshwater, is a qualitatively different substance due to the presence of dissolved mineral salts in it. Seawater to dilute electrolyte solutions is dissociated into ions: cations (hydrogen and metal ions) and anions (acidic and aqueous residues). Studies of the number and composition of phases during freezing of seawater have shown that a certain temperature corresponds to a certain concentration of brine in the system and interphase physicochemical reactions; moreover, the chemical composition of the entire system as a whole should remain constant. The sequence and temperature of the transition of basic salts from seawater to the solid phase have been experimentally established. These data are presented in Table 3.1.

At a temperature of $-22.6\,°C$, the bulk of the salts is in solution. Figure 3.2 shows a phase diagram of the dependence of the mass ratios of pure ice, brine, and the total mass of salt crystallohydrates, as well as their ions on the ice temperature. The figure shows at what temperature the crystallization of salts $CaCO_3 \cdot 6H_2O$, $Na_2SO_4 \cdot 10H_2O$, etc. occurs. Using the diagram, we can determine the amount of pure ice, brine, and precipitated salts.

The relative proportion of fresh ice in sea ice can be determined from the ratio [14]

Table 3.1 Experimental data on the sequence and temperature of the transition of basic salts from seawater to the solid phase

Solid phase	Temperature of the system (°C)
Ice	−1.8
$CaCO_3 \cdot 6H_2O$	−1.9
$Na_2SO_4 \cdot 10H_2O$	−7.6, …, −8.2
$CaCO_4 \cdot 2H_2O$	−15.0
$NaCl \cdot 2H_2O$	−22.6
KCl	−34.2
$MgCl_2 \cdot 12H_2O$	−33.6, …, −36.0
$CaCl_2 \cdot 6H_2O$	−55.0

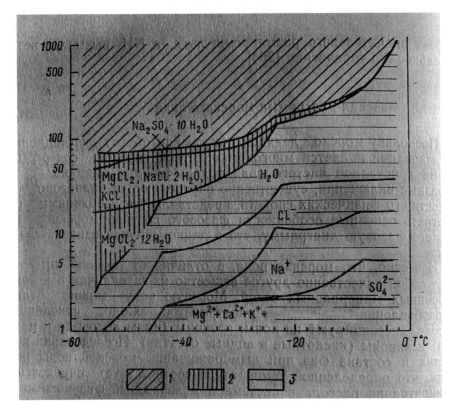

Fig. 3.2 Sea ice phase diagram

$$\frac{m_f}{m} = 1 - \frac{S_i}{S_b} + \frac{m_{cs}}{mS_b}(1 - S_b) \tag{3.1}$$

where m_f—mass of fresh ice, m—the total mass, which is the sum of the masses of fresh ice, liquid water and salts; m_{cs}—the mass of salts in the crystalline state; S_i—salinity of ice, S_b—brine salinity.

3.2 Structural and Genetic Classification of Ice

Currently, there are a large number of classifications of ice. This is due to the fact that their authors proceeded from different principles of division. So, for example, based on the requests of ice engineering, ice can be divided into eight classes [10, 15]:

(1) atmospheric ice (snow, frost, hail, ice), (2) surface ice of water areas (ice, covering oceans, seas, rivers, lakes and other reservoirs in winter), (3) intrawater ice (bottom ice, sludge), (4) continental ice (various kinds of glaciers, icebergs,

ice islands, (5) permafrost ice (which are formed in the lithosphere), (6) buried ice (remnants of continental ice, deposited by marine sediments), (7) ice of special formations (ice, formed from foam and spray, thrown by the wind on the shores; vein ice, formed in soils and rocks in the form of veins), (8) ice, artificially created by man (low-temperature amorphous ice, obtained by condensation of steam, supercooled to $-120, ..., -140$ °C; «hot» and «heavy» ice—modifications of ice, having a higher density than water and existing at positive temperature and high pressure).

The above classification, however, does not show much on the structural features of ice and the conditions of their formation. For example, Arctic and Antarctic sea ice was usually divided by age, distinguishing perennial, annual, autumn, winter ice, and sometimes drifting or soldered. In freshwater bodies, ice is divided into river and lake, and according to the conditions of their formation—into water, sludge, or snow [16]. Sometimes the terms «black ice» and «white ice» are found. «Black» is ice, formed when water freezes with a small amount of scattering inclusions; in the array, such ice has a dark, almost black color. «White ice» is formed when sludge or snow freezes with a large number of air inclusions of a fine-crystalline structure. In the array, due to light scattering, the ice has a white color.

Classification of ice only by structural features (size, orientation, and shape of crystals) implies the division of natural ice into prismatic, fibrous, and isometric (granular). At the same time, each type of structure is divided into correctly granular (hypidiomorphic) and incorrectly granular (allotriomorphic) by the nature of the development of faces (complexity of surfaces), and by the size of crystals—into small, medium, and coarse-grained. Genetic classification of sea ice by V.H. Buinitsky differentiates ice only according to the conditions of their formation, distinguishing at the same time conjugation, conjugation-intrawater, infiltration, etc. Structural features are not taken into account in this classification. In 1969, Michel and Ramseier proposed a genetic classification, according to which freshwater ice is divided into three main groups and seven types. This classification is often referred to by researchers. All the mentioned classifications have some disadvantages. Below we consider the classification, proposed by N.V. Cherepanov and widely used recently in the Russian Federation, as well as briefly dwell on the most recent classification, currently being developed by the Working group of the Ice Committee of the IAHR (International Association for Hydraulic Research).

3.2.1 Classification of Ice of Natural Reservoirs by N.V. Cherepanov. Possibility of Use for Aviation

The classification, proposed by N.V. Cherepanov, takes into account the following main genetic (conditions of formation) and structural (crystal structure) features [17]:

(1) salt composition of reservoirs, their division into freshwater, desalinated, and saltwater;

(2) hydrometeorological regime of reservoirs following the crystallization of water, and the period of ice build-up (the magnitude and direction of the temperature gradient, convective mixing, currents, drift, the possibility of formation of intrawater ice);

(3) features of the growth of ice crystals (manifestation of the law of geometric selection, the influence of a natural magnetic field);

(4) morphological characteristics of crystals (shape of growth and size of crystals, degree of perfection of crystal faces and their orientation, characteristics of inclusions);

(5) the duration of the existence of the ice cover and the processes of metamorphic transformation, associated with it: thermal metamorphism, associated with changes in the structure of ice under the influence of selective radiation melting, refreezing, infiltration, firnization, etc. and dynamic metamorphism, associated with various kinds of violations of the integrity of crystals as a result of polygonization of stressed crystals and cataclastic phenomena.

According to the condition of ice formation and salinity of water, Cherepanov divides the ice of natural reservoirs into four main groups. The first group (group A) includes ice, forming in fresh or highly desalinated (Si < 2%) reservoirs. The second group (group B) includes ice, formed in desalinated reservoirs (Si = 2–24.7%). The third group (group C) includes sea reservoirs (Si > 24.7%). The fourth group (group D) includes ice, formed as a result of the metamorphic transformation of a long-existing ice sheet, that has undergone thermal and dynamic metamorphism. Ice groups are divided into growth types, which are indicated by digital indexes (for example, A1, A2, etc.).

Table 3.2 and Fig. 3.3 give the characteristics of the main types of ice crystal growth, depending on the magnitude and direction of the temperature gradient at the water–ice partition surface.

For the first type of growth (Fig. 3.3a), the temperature gradient in the near-ice water layer is 1–4 °C/m, and the isothermal surface of the crystallization front corresponds to the freezing temperature of water [18, 19]. The constraint of crystal growth and geometric selection leads to the development of crystals in the direction of the C-axis. With an average diameter of the crystals of the primary ice layer of 2–3 mm and a temperature gradient of 1.5 °C/m, a complete adjustment in orientation occurs at the horizons of 10–15 mm, with a gradient of 0.4 °C/m, the boundary of

Table 3.2 Characteristics of the main types of ice crystal growth

Type of crystal growth	Temperature gradient	Growth rate of crystals	Crystallization front	Ice–water interface
The first	Big	Small	Flat	Smooth
The second	Small	Average	Narrow zone	Rough
The third	Missing or negative	Big	Wide area	Openwork layer

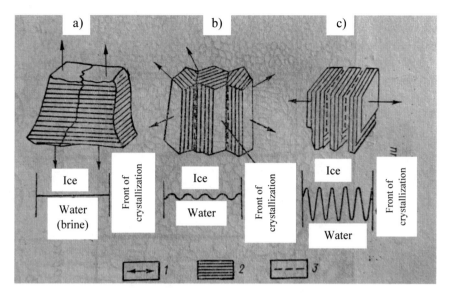

Fig. 3.3 Three types of crystal growth

the restructuring of the structure from chaotically oriented crystals to ordered ones ends at the horizons of 25–40 mm.

For the second type of growth (Fig. 3.3b), the temperature gradient is equal to tenths-hundredths of °C/m, and the front of crystallization zone is narrow [20–22]. A peculiar roughness appears on the lower surface of the ice due to the uneven growth rate of differently oriented crystals. Under such conditions, crystals are usually formed in the form of pointed pyramids, prisms, poles, and needles. The C-axes of crystals do not have an ordered orientation, but sometimes their orientation is close to horizontal. This is usually observed at a horizon of about 5 cm at a temperature gradient of 0.01 °C/m.

The third type of growth (Fig. 3.3c) is observed in the absence of a temperature gradient in the near-ice water layer (often accompanied by a slight supercooling of the water) [23]. This type is characterized by the highest crystallization rate, skeletal crystals grow in the direction of the base plane. Such conditions in the ice layer are created only in sea reservoirs with well-developed vertical mixing, providing a stable homogeneous state of the ice layer of water with a temperature, equal to the freezing point. In sea ice, the size of the crystallization zone is indicated by the presence of an openwork layer at its lower surface.

The peculiarity of the construction of the Cherepanov classification is that the types of ice in each group are arranged «according to an increasing digital index, which corresponds to an increase in the dynamism of the reservoir, a violation of its thermal regime, an increase in the role of intrawater ice». The general scheme of classification of freshwater and salt ice, according to Cherepanov, is shown in Fig. 3.4. It also provides a schematic representation of the types of ice in the horizontal and

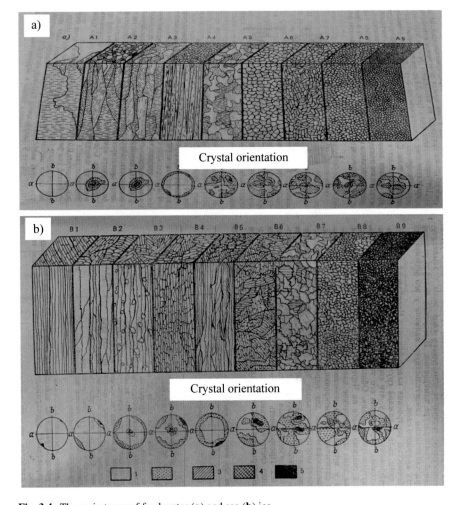

Fig. 3.4 The main types of freshwater (**a**) and sea (**b**) ice

vertical planes, provides typical stereograms of optical orientation, geometric models of crystals, used to determine the coefficient of curvature of the crystal surface, when calculating the surface of a single crystal and the total surface of crystals in a unit volume. In order to quantitatively describe the textural features and transparency of ice, the scale, presented in Table 3.3, was used. The transparency of ice was visually assessed by the degree of attenuation of light by plates 10 cm thick. In Table. 3.4, some physical characteristics of various types of ice are given.

Types of freshwater ice:

A1 is common on small lakes, lagoons, small unregulated reservoirs with a sufficiently large thermal reserve;

Table 3.3 Scale for visual assessment of the number of inclusions in the ice

Number	General characteristics of inclusions	Symbols, points
1	Ice has no inclusions or inclusions, larger than 1 mm, are single, no more than 1–10 per dm^2	0
2	Rare inclusions, more than 10 by 1 dm^2, do not change the overall transparency of the ice; small print is freely read through the plate	1
3	A small number of inclusions, only large capital letters are freely distinguished	2
4	The average number of inclusions, silhouettes of large (up to 1 cm) letters are viewed	3
5	A large number of inclusions, silhouettes of large letters are not visible, only the contours of large objects differ	4
6	A very large number of inclusions, the ice is completely non-transparent	5

Table 3.4 Physical characteristics of different types of ice

Type of ice	Salinity ($^0/_{00}$)	The number of inclusions in the ice, according to the 5-point system	Ice density (kg/m^3)	Prevailing orientation C-axis	Average crystal sizes (mm)			The coefficient of tortuosity of crystals
					a	b	c	
A1	–	0	915–917	Vertical	250	250	500	1,70
A2	–	1	910–917	Mixed	120	120	220	1,55
A3	–	2	900–915	Chaotic	95	95	95	1,50
A4	–	2	905–917	Horizontal	35	35	500	1,20
A5	–	3	880–900	Chaotic	32	32	45	1,60
A6	–	4	830–900	Chaotic	8	8	8	1,45
A7	–	4	820–850	Chaotic	3	3	5	1,10
A8	–	5	810–850	Chaotic	2,5	2,5	2,5	1,15
A9	–		790–820	Chaotic	1.5	1.5	1.5	1.30
C1	1.5–4.0	3	910–930	Horizontal	0.7	4	100	1.10
C2	1.5–4.0	3	910–930	Horizontal	0.6	4	100	1.15
C3	1.6–3.5	3	930–950	Horizontal	0.7	4	50	1.25
	1.6–3.5		930–950	Vertical	5	5	5	1.10
C4	1.6–3.0	4	900–920	Mixed	0.6	3	20	1.12
C5	4.0–15.0	3	900–935	Waist	0.4	3	20	1.10
C6	2.0–5.5	3	890–900	Chaotic	10	10	15	1.40
C7	2.0–6.0	4	880–900	Chaotic	7	7	10	1.35

(continued)

Table 3.4 (continued)

Type of ice	Salinity ($^0/_{00}$)	The number of inclusions in the ice, according to the 5-point system	Ice density (kg/m^3)	Prevailing orientation C-axis	Average crystal sizes (mm)			The coefficient of tortuosity of crystals
					a	b	c	
C8	2.0–7.0	5	850–900	Chaotic	3	3	4	1.25
C9	2.0–14.0	5	830–900	Chaotic	1.2	1.2	1.2	1.15
D1	1.4	3	910–915	Relict	10	10	14	1.80
D2	0.5	2	890–905	Relict (waist)	1.5	10	200	1.30
D3	0.1	2	870–900	Chaotic	12	12	12	1.90
D4	0.0	3	760–820	Chaotic	3	3	3	1.25

A2 is most widespread on large open lakes, reservoirs, large rivers after prolonged wind mixing, accompanied by significant heat storage;

A3 is formed on regulated rivers, reservoirs, large lakes, which are characterized by a rhythmic violation of the temperature conditions of the ice layer and where non-freezing areas of water remain for a long time;

A4 is formed on all freshwater bodies, which by the time of ice formation lose all heat reserves. Such ice is common in the mouths of rivers, flowing into the Arctic seas, and is formed, when the surface of desalinated layers of the sea freeze with a large accumulation of meltwater after the summer melting of sea ice;

A5 conditions for the formation of this type of ice are created on regulated rivers, reservoirs, large lakes with a complex hydrological regime, characterized by a continuous violation of temperature stratification;

A6 is formed in reservoirs with steady cooling of the ice layer of water to the freezing point or its slight overcooling, contributing to the emergence of new crystals, that disrupt the oriented growth of crystals. Such ice formation occurs more often on fast-flowing rivers, where wormwoods persist for a long time;

A7 is typical for deep-water sections of rivers with the presence of open water areas (wormwood). Such ice is formed from crystals of intrawater ice, developing in the water column;

A8 is formed, when dense accumulations of sludge and snow in autumn freeze with intense wind mixing of surface water layers, for example, on rivers with a fast current;

A9 is formed, when snow, soaked with water, freezes, for example, during the formation of various kinds of ice.

The structure of Group B ice is still poorly understood, so there is no fractional typing yet.

The ice of sea reservoirs, belonging to group C, is also divided into nine types:

C1 is formed in a stable layer with well-developed vertical convection. It is found in the solder outside the zone of distribution of mainland river water runoff. This type of ice has a well-defined spatial ordering of the horizontally arranged C-axes of crystal fibers;

C2 is the most widespread; it is mainly vast fields of drifting ice in the Arctic and part of the soldered ice in the Antarctic;

C3 is formed, when the stable growth of crystals is disrupted and the elements of intrawater ice enter the front of crystallization intensively. This type is typical for drifting and soldered Arctic ice in areas of intense movement and the presence of stationary wormwoods, respectively;

C4 differs little from type C3 and is found in the same place;

C5 is formed during intensive freezing of seawater in winter in the presence of a large number of crystallization centers on the surface of the reservoir;

C6 appears during the freezing of sea reservoirs with a sharply changing hydro-logical regime—large changes in the speed and direction of currents, sharp fluctuations in water temperature, salinity of seas, straits and areas, adjacent to river mouths;

C7 is formed by strong wind mixing of shallow areas of the sea, accompanied by intensive formation of sludge from intrawater and bottom ice;

C8 is formed by freezing of snow grains with small grains of intrawater ice in seawater. It is widespread everywhere in the seas, especially widely in Antarctica;

C9 is formed in the process of gradual infiltration of seawater, protruding to the ice surface under the weight of snow and ice cover.

When using ice surfaces for the construction of ice airfields, it is necessary to know the structure of ice in a particular region, as well as its variability, taking into account the peculiarities of meteorological phenomena and water properties. This is of funda-mental importance for forecasting not only the thickness of the ice, depending on the season, but also its strength characteristics. Such a circumstance is of crucial importance, when landing large aircraft and long-term use of ice airfields.

3.2.2   General Classification of Ice, Proposed by the Ice Committee of the International Association for Hydraulic Research

Taking into account the non-universality of some and the complexity of others from among the well-known analyzed classifications, the Working Group of the Ice Committee of the IAHR (International Association for Hydraulic Research) devel-oped and sent out to a wide range of specialists for discussion and amendment a general classification, according to which it was proposed to designate all types of natural ice with a code, consisting of letters and numbers [24–26]. The classification scheme, designed for simplified identification of ice types by sections, takes into account five structural, textural, and other characteristics: (I) type of grains: A—shapeless grains; B—in the form of blocks of groups of homogeneous grains; C—columnar grains; D—disk-shaped; E—equiaxed and F—fibrous grains; (II) orientation of crystallographic axes: 0—chaotic; 1—predominantly vertical; 2—predominantly in the horizontal plane; 3—predominantly ordered in the horizontal plane; 4—indeterminate; (III) inclusions: a—air; b—brine; c—solid inclusions, soil, algae; (IV) origin: RL—river and lake; 1G—icebergs and shelf glaciers; FY—annual sea ice; MY—long-term sea ice; A—ice of atmospheric origin; (V) morphology: P—drifting ice; R—ridge of hummocks; S—relatively flat ice fields of drifting or soldered ice.

For example, the code of an ice sample from a ridge of hummocks of a sea long-term ice cover, characterized by grains of indeterminate shape and chaotic orientation of the C-axes, as well as the presence of brine, is written as A0MYR.

3.3 The Main Characteristics of the Physical State of Sea Ice and Their Seasonal Variability

It is known that the electrical (dielectric permittivity, absorption, and scattering of electromagnetic waves) and physico-mechanical (elasticity, plasticity, strength) properties of ice are determined by such characteristics of its physical state as, for example, density and porosity, propagation velocities of electromagnetic and elastic waves, which primarily depend on the crystal structure of ice, its structure, temperature and salinity [27]. Note also that knowledge of the temperature and salinity of ice gives us the opportunity to quantify another important characteristic of sea ice for practical purposes—the relative content of the liquid phase in it in the form of brine inclusions.

Currently, a large number of partial dependences of some of the most applicable physical and mechanical characteristics of ice in engineering calculations on its temperature T, salinity S_i, crystal sizes d_{er}, density p, etc. from simple empirical formulas have been obtained. These formulas connect two or three parameters (for example, the formulas of Gold, Berdennikov, Tabata, Pounder, etc. for the Young modulus in the form of $E = f(T)$, $E = f(T, S_i)$, $E = f(\rho)$… or the formulas of Frankenstein, Ryvlin, Peyton for the strength of ice for bending of the type $\sigma_{bend} = f(S_i)$, $\sigma_{bend} = f(C)$, etc. up to quite complex (with many parameters) phenomenological equations of ice state. As an example, we give the Zaretsky equation with five parameters, proposed to determine the deformation of ice at any given time for a given, in particular, constant uniaxial load, and, consequently, for the possibility of determining the modulus of deformation and strength of ice as a function of time.

Using these dependencies, in principle, it is possible to estimate the seasonal variability of the physical quantities of ice, that characterize its behavior under load, including, for example, the bearing capacity of the ice cover.

Currently, due to the growing needs of engineering practice (which, of course, is connected with the organization of transportation in winter conditions and the development of hydrocarbon deposits on the Arctic coast, etc.), the demand for generalizations of theoretical and experimental studies of the physical properties of ice, obtained in recent years, has significantly increased. It is obvious, that such generalizations are of particular value, which would allow us to perform a numerical (and if there are sufficient series of observations, probabilistic) analysis of the seasonal variability of some characteristics of sea ice, the improvement of knowledge about which we need both to improve the assessment of the parameters of the sea ice cover and to solve the problems of protecting engineering structures from the effects of ice [28–30].

However, both the first and the second are closely related. It should also be noted that the analysis of the seasonal variability of these characteristics should preferably be carried out with reference to the types of ice (or its age gradations) in accordance with some generally accepted classification and nomenclature of sea ice WMO so, that it is possible, using ice atlases, to proceed to a regional statistical assessment of the physical properties of ice. Below are the results of generalizations of calculated and experimental data on the temperature, density, porosity, and salinity of sea ice for the following six age gradations: young ice with a thickness of 10–30 cm, thin annual 30–70 cm, annual ice with an average thickness of 70–120 cm, thick 120–200 cm, two-year ice 200–300 cm and long-term ice with a thickness of more than 300 cm.

3.3.1 Ice Temperature

Mathematical methods for calculating the temperature regime of natural ice are currently quite well developed [24, 27, 31–35].

Table 3.5 shows the results of solving the problem of changing the temperature of freshwater ice at various horizons in 10 h with the following initial data: snow thickness 0.5 cm; ice thickness 60 cm; thermal conductivity of snow 0.56 W/(m · K); the temperature conductivity of snow 3.5×10^{-7} m^2/s; thermal conductivity of ice—0.66 W/(m · K); the temperature conductivity of ice is 1.4×10^{-7} m^2/s. From the data in this table, it can be seen that the surface temperature of ice in contact with snow for 3 h, regardless of the change in air temperature, remains almost constant, equal to $-15\,°C$. Over the next 7 h, it rises from $-14.23\,°C$ to $-11.72\,°C$. The data of field observations showed that three types of temperature distribution curves are characteristic of the ice cover of freshwater water areas. The first type has a positive gradient and is typical for periods of cooling; it is more common at the beginning of winter. The second type has a minimum temperature in the middle layers of the ice sheet and is characteristic of the period of increasing air temperature; it is more common in spring. The third type is transitional from the first to the second. It is characterized by the presence of two temperature minima. It can occur throughout the entire winter period.

Table 3.5 Calculated temperature values in the snow-ice cover of a freshwater reservoir, when the air temperature changes from −30 to −14,16 °C

Distance from the surface (cm)	Time (h)										
	0	1	2	3	4	5	6	7	8	9	10
Snow											
0	−30	−25.45	−23.02	−21.55	−21.01	−18.86	−17.17	−16.20	−15.40	−14.80	−14.16
5	−25	−25	−22.72	−21.51	−20.21	−19.04	−17.90	−16.80	−15.92	−15.27	−14.57
10	−25	−20	−20	−18.86	−18.07	−17.22	−16.43	−15.64	−14.93	−14.34	−13.70
15	−15	−15	−15	−14.62	−14.23	−13.82	−13.38	−13.07	−12.56	−12.13	−11.72
Ice											
25	−12.5	−12.5	−12.5	−12.5	−12.31	−12.12	−11.85	−11.60	−11.37	−11.03	−10.73
35	−10	−10	−10	−10	−10	−9.9	−9.85	−9.65	−9.50	−9.34	−9.13
45	−7.5	−7.5	−7.5	−7.5	−7.5	−7.45	−7.45	−7.40	−7.31	−7.24	−7.12
55	−5	−5	−5	−5	−5	−5	−5	−4.97	−4.95	−4.9	−4.85
65	−2.5	−2.5	−2.5	−2.5	−2.5	−2.5	−2.5	−2.5	−2.48	−2.47	−2.44
75	0	0	0	0	0	0	0	0	0	0	0

The main features of the vertical profile of sea ice temperature and its changes over time are reduced by the fact that the temperature of the lower surface of the ice is close to the freezing temperature, and the upper surface is close to the air temperature. For naturally occurring changes in air temperature, the vertical temperature profile of thin ice differs relatively slightly from the linear one [36]. As the thickness of the ice increases, its heat capacity increases and the delay in temperature changes with depth becomes more noticeable. It has been shown that long-term ice is characterized by a delay in the extremum of the average monthly temperature with depth, as a result of which its vertical profile differs from the linear one.

The graph for typing the temperature distribution in the thickness of the long-term sea ice sheet, according to experimental data, for calculating the thickness-weighted average and monthly average temperatures of $\tilde{T}_i$ is shown in Fig. 3.5. For the same purpose, other data from field studies of the temperature regime of the ice sheet of various thicknesses were used. The average monthly values of the air temperature of the T_a as typical for the middle latitudes of the Arctic, obtained from observations of seasonal variability of ice temperature in various layers of relatively thick (h = 1.6 m) ice cover, provided, that the surface temperature of thin young ice does not coincide with the measured air temperature, were taken as the basis for estimating the $\tilde{T}_i$ of the six selected age gradations at standard height usually. At the same time, the calculation was carried out according to the following formulas [37–40].

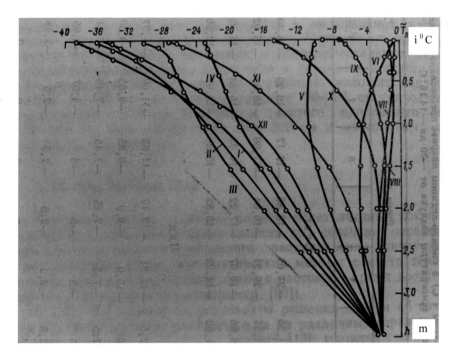

Fig. 3.5 The change in the average monthly temperature by the thickness of the long-term sea ice

for ice $h > 50$ cm:

$$\widetilde{T}_i = \frac{1}{h} \sum_{i=1}^{n} \overline{T}_{av} \cdot h_i \tag{3.2}$$

where $\overline{T}_{av}$—average temperature of the i-th layer by thickness Δh_i;

for ice $h < 50$ cm (assuming the linearity of the temperature profile over the entire thickness)

$$|\widetilde{T}_i| = \frac{1}{2}(|\widetilde{T}_a| - \Delta T - 2) + 2 \tag{3.3}$$

where ΔT—the difference between the surface temperature of young ice and air at standard altitude. Since ΔT is a value, that depends on air temperature and ice thickness, its values were taken from a graph, based on experimental data from Makshtas and their linear extrapolation to calculate the thickness of ice h (cm) by the number of degree days of «frost».

The family of six curves in Fig. 3.6 gives an idea of the seasonal variability of the weighted average temperature of the ice sheets of the selected age gradations for the middle latitudes of the Arctic.

3.3.2 Salinity

Salinity strongly depends on meteorological conditions during the formation of the ice sheet. At a low temperature, the crystal growth rate is higher and their sizes are smaller than at elevated temperatures, which is the reason for the large capture of liquid brine in the intercrystalline layers [41].

Brine flows out less, when ice grows during wind waves, which together with low temperature creates conditions for the formation of ice with increased salinity. If snow also falls at the same time, then the salinity of the initial forms of ice with a thickness of about 10 cm can reach values only two times less than the salinity of the water, from which the ice was formed. For the Arctic conditions, noted above, the maximum salinity of the ice is 17‰.

At the same time, at a temperature of −40 °C and the absence of wind, higher salinity values of the initial forms of ice were also noted: 2 h after the beginning of formation, 26‰, after 12 h, 19‰ and after a day, 16‰.

With an increase in the thickness of the ice, its growth rate becomes less, and the crystals are larger, which contributes to the outflow of brine. However, this decrease goes only to some layers, in which the capillary rise from the water-saturated lower layers does not affect [42–44].

The most rapid process of desalination of the upper layers occurs during the summer melting.

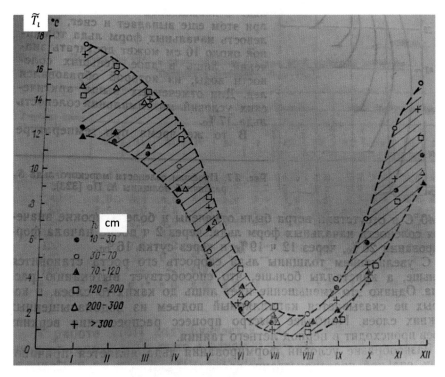

Fig. 3.6 Seasonal variability of the average monthly temperature of the ice sheet of different thickness

The variety of ice formation conditions is the reason for the very strong spatial and temporal variability of the absolute salinity values of sea ice. Naturally, the ice of a more temperate climatic zone has less salinity than, for example, the ice of the high-latitude Arctic. Typical distributions of sea ice by salinity thickness are shown in Fig. 3.7 and in Table 3.6.

Some works [45–50] emphasize that the actual salinity of ice from year to year, even in the same area, may differ greatly from its average value. Therefore, measurements at a number of points in a small area can give a salinity dispersion of the same order of magnitude as its average value.

3.3.3   Density

The density of sea ice varies significantly from season to season and significantly depends on temperature and salinity [51]. We present data, according to which the density of winter ice is 860–920 kg/m^3 for annual ice and 830–900 kg/m^3 for long-term ice. In summer, the density of strongly weathered ice due to brine runoff drops to

Fig. 3.7 Salinity profiles of
sea ice of various thicknesses

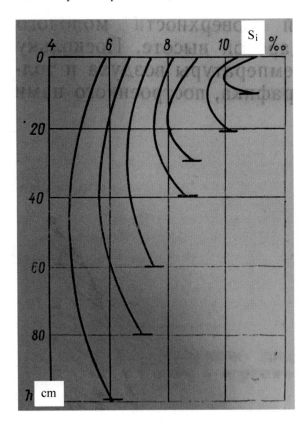

Table 3.6 Layered and weighted average values of sea ice salinity for the entire thickness

Layer depth (cm)	Island privacy				Vilkitsky strait				Matochkin sphere strait			
	XII	II	III	IV	XII	I	III	IV	XII	II	III	IV
0	9.4	6.9	8.6	8.6	6.5	4.6	4.6	4.6	9.3	7.5	9.4	6.8
20	6.6	4.4	5.2	4.1	2.7	3.7	3.9	4.2	6.9	6.7	5.8	6.4
40	4.9	5.4	4.8	4.6	2.6	3.4	3.6	3.7	6.3	5.8	6.4	6.1
60	3.5	4.0	4.5	4.7	3.0	2.8	3.0	3.8	10.4	4.0	5.0	5.1
80	6.5	4.7	4.9	4.5	2.4	3.1	3.0	3.1	–	–	4.2	4.2
100	–	4.9	4.0	5.0	–	3.8	4.3	4.1	–	5.5	5.8	5.1
120	–	7.6	3.2	4.4	–	–	–	3.7	–	–	4.2	6.0
140	–	–	9.0	5.7	–	–	–	4.1	–	–	–	–
160	–	–	–	4.5	–	–	–	–	–	–	–	–
S_i	5.75	5.1	5.0	4.9	3.0	3.4	3.3	3.8	7.7	5.75	5.7	5.55

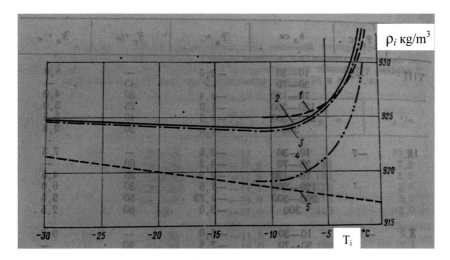

Fig. 3.8 Dependence of the density of columnar sea ice on salinity $5^0/_{00}$ from temperature T_i, according to data of various authors. 1—Zubov, 2—Cox and Weeks, 3—Anderson, 4—Shverdfeger, 5—Pounder data for freshwater ice

560–640 kg/m³, and long-term ice has a strong spread: 620–933 kg/m³. The middle layers of the ice cover have the highest density, the upper and lower ones have the lowest density.

The calculated values of the density (excluding air inclusions) of sea ice can be taken as reference points, when analyzing the seasonal variability of the available experimental data on ice density of various age gradations. Figure 3.8 shows the dependence of the density of sea ice with a salinity of 5‰ on temperature, according to data of various authors.

In some papers, the results are presented in the form of histograms of distributions of the values of the density of ρ_a and the relative volume of the air pores of υ_a, as well as the regression dependence of $\upsilon_a = f\ (\rho_a)$. The highest density values are observed in young ice. However, averaging over the entire series of observations for these ices gives only a figure of about 915 кg/m³, while for the January nilas near the Labrador peninsula, a value of 945 кg/m³ is given. In the Kara Sea and the Laptev Sea, the maximum value was 923 кg/m³. During the winter and during the transition of ice from nilas to annual medium and annual thick, the density changes little.

3.3.4   Relative Content of Gas and Brine in Sea Ice Samples

The total porosity of sea ice, taking into account the relative content of both gas and brine per unit volume, undoubtedly plays an important role in controlling its properties [52, 53]. Note, that with a relative volume of the liquid phase, equal to 0.10, the properties of ice change in the following ratio: the coefficient of thermal conductivity—by 10%, the bending strength—by more than 2 times, the modulus of elasticity—by 4 times, etc. This conclusion is also reached by considering a cellular model of the structure of sea ice, various modifications of which are used to theoretically describe, in particular, the mechanical and electrical properties of ice.

As you can see, the combined effect of the air and brine, contained in the ice, on its properties can be significant, and in some cases large. So, if in winter the total volume of the gaseous and liquid phases in the ice is usually units of percent, then during the summer melting it reaches 20% or more. However, even in winter, the total air volume of individual layers of ice cover and brine can be higher than 10%.

However, the total porosity is rarely measured. Usually mechanical, thermophysical, and electrical properties are represented as dependencies on the volume of brine. There are several reasons for neglecting the second component of porosity (gas volume) [54]: (1) direct measurement of gas volume takes a lot of time, (2) it is not so easy to calculate the volume of gas in natural conditions, and (3) most of the studies of sea ice focused on annual ice, in which it is assumed, that the volume of gas is relatively small, compared to the volume of brine, and therefore it (the volume of gas) in the first approximation could be neglected. Such an approximation, however, would be clearly unsatisfactory in relation, for example, to long-term ice, from which most of the brine drains (flows out) and is replaced by gas (most often, apparently, air).

The relative volume υ_a of gas pores in ice (or its porosity) is related to the density by the ratio $\upsilon_a = 1 - \rho_a/\rho_0$, where ρ_0 is the density of ice, in which there are no gas pores, and ρ_0 is the density of ice, containing pores with air, the relative volume of which is equal to υ_a. To calculate the υ_a, data from full-scale definitions of the ρ_a, temperature T and salinity S_i of ice are used, and the values of the ρ_0 are calculated from known T and S_i.

Since most experimental determinations were performed at $T = -2, \dots, -10\,°C$, at which the effect of the brine, contained in the ice, on the density is much less than the effect of air pores, there is a close quantitative relationship between ρ_a and υ_a, despite the difference in the salinity of the ice. The graph, shown in Fig. 3.9, can be used to approximate the estimate of a υ_a by ρ_a.

Moreover, when statistically processing large series of sea ice porosity values, it seems possible to indirectly determine its salinity, provided, of course, that the temperature of the ice is known [55].

Table 3.7 shows the results of calculating the υ_a for various age gradations of ice.

It follows from Table 3.7 that the largest volume of pores was observed in the fields of gray ice. For other age forms of ice, the average values relative to the volume of air pores turned out to be of the same order 0.009–0.013. For ice 0.3–0.7 m thick with a probability of 62% the values $\upsilon_a = 0.005$–0.015, and for ice 0.7–2.5 m thick in 85 cases out of 100, the values $\upsilon_a = 0.002$–0.020 are almost equally likely. The calculation of the average values for individual layers (upper, middle, and lower) gave the following numerical characteristics [56, 57]:

annual ice—0,0184; 0,0083;0,0120;
young ice—0,0153; 0,0125; 0,0104.

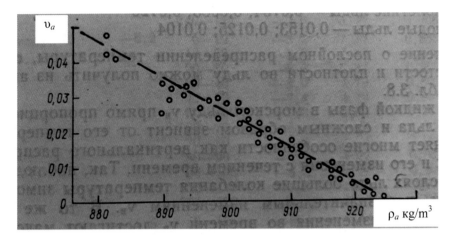

Fig. 3.9 Dependence of the density of Arctic ice on the relative volume of air pores, contained in it

Table 3.7 Porosity of ice of various ages

Age of ice	Porosity of ice			Number of vertical profiles
	Average	Maximum	Minimum	
Nilas (autumn)	0.0088	0.0482	0.0046	23
Grey (autumn)	0.0407	0.0690	0.0121	24
Annual				
Thin (winter-spring)	0.0136	0.0274	0.0030	16
Medium and thick (early spring)	0.0133	0.0333	0.0026	49

An idea of the layer-by-layer distribution of temperature, salinity, porosity, and density in ice can be obtained from the data analysis in Table 3.8.

The volume of the liquid phase in sea ice v_v is directly proportional to the salinity of the ice and depends in a complex way on its temperature, which explains many features of both the vertical distribution v_v and its changes over time. Thus, the large temperature fluctuations, observed in the upper layers of ice in winter, lead to only minor changes in v_v. At the same time, in the middle layers, the changes in time v_v reach a maximum due to the fact, that the temperature of the ice is close to its melting point (this is especially true for ice near the eutectic points -8.2 and $-22.6\ °C$).

Currently, to determine the content of brine in ice, one can use, for example, the Frankenstein-Garner formula [56–59]:

$$v_v = \frac{S_i}{1000}\left(\frac{-49,185}{T} + 0,532\right) \tag{3.4}$$

Table 3.8 Experimental characteristics of various types of ice in the Kara Sea

Type of ice	Air temperature (°C)	Thickness layer of ice (cm)	Layer temperature (°C)	Salinity of ice in the layer	The density of ice in the layer (g/cm^3)	Porosity coefficient
Drifting annual	−12	0–10	−13.2	7.42	0.884	4.6
		10–20	−11.6	5.21	0.904	2.2
		20–30	−10.6	4.98	0.910	1.6
		30–40	−9.6	4.98	0.910	1.6
		40–50	−8.8	4.98	0.913	1.3
		50–60	−7.8	4.20	0.917	0.8
		60–70	−7.2	4.20	0.917	0.8
		70–80	−6.4	4.80	0.917	0.8
		80–90	−5.8	4.56	0.920	0.6
		90–100	−4.6	4.56	0.920	0.6
		100–110	−3.8	4.56	0.917	0.9
		110–120	−3.4	4.45	0.917	0.9
		120–130	−2.6	5.57	0.908	1.9
		130–140	−1.6			
	−11	0–10	−11.5	9.15	0.881	4.8
		10–20	−11.6	6.04	0.900	2.6
		20–30	−11.6	5.1	0.903	2.2
		30–40	−10.8	5.1	0.903	2.2
		40–50	−10.2	5.1	0.904	2.1
		50–60	−9.4	4.51	0.904	2.1
		60–70	−9.2	4.51	0.904	2.1
		70–80	−8.8	4.51	0.909	1.6
		80–90	−7.0	4.38	0.909	1.6
		90–100	−6.6	4.38	0.909	1.6
		100–110	−5.8	4.38	0.909	1.6
		110–120	−4.6	4.34	0.911	1.4
		120–130	−4.8	4.34	0.911	1.4
		130–140	−2.8	5.08	0.906	1.9
		140–148	−2.0	5.08	0.906	1.9
One-year fat	−6.2	0–10	−8.0	7.70	0.855	7.4
		10–20	−8.0	7.11	0.852	7.5
		20–30	−8.0	5.37	0.892	3.6
		30–40	−8.0	4.70	0.893	3.6
		40–50	−7.4	4.70	0.904	2.3

(continued)

Table 3.8 (continued)

Type of ice	Air temperature (°C)	Thickness layer of ice (cm)	Layer temperature (°C)	Salinity of ice in the layer	The density of ice in the layer (g/cm^3)	Porosity coefficient
		50–60	−7.0	3.84	0.904	2.3
		60–70	−6.4	3.84	0.904	2.3
		70–80	−5.2	3.84	0.904	2.3
		80–90	−4.8	4.22	0.908	1.9
		90–100	−3.8	4.22	0.908	1.9
		100–110	−3.2	4.22	0.919	0.8
		110–120	−2.2	3.73	0.919	0.8
Soldered	−11.6	0–10	−6.4	5.23	0.893	3.3
		10–20	−6.8	4.02	0.911	3.0
		20–30	−6.6	4.15	0.913	1.4
		30–40	−6.0	4.15	0.913	1.4
		40–50	−5.0	3.62	0.893	1.2
		50–60	−4.2	3.62	0.893	1.2
		60–70	−3.2	3.28	0.909	1.7
		70–84	−2.4	3.28	0.909	1.7

(T—temperature, °C (T < 0); S_i—salinity, ‰, and v_v—liquid phase in relative units), which, being an empirical dependence for ice temperatures of −0.5, …, − 22.9 °C, has found wide practical application. The results of calculated estimates and generalization of experimental data are presented in Fig. 3.10 in the form of seasonal variability of the presence of brine in sea ice, for selected six age gradations.

3.4 Thermophysical Properties

The formation and destruction of the sea ice cover is an energy process, characterized by the absorption of heat or the release of a large amount of it. This process largely determines the energy balance of atmospheric circulation on the scale of the entire Earth [31]. The ice sheet, being a product of the interaction of the ocean and the atmosphere, is one of the very indicative indicators of climate change on our planet, and acts as an intermediary in the thermal and dynamic interaction of the atmosphere and the ocean. The latter circumstance seems to be the most important incentive to replenish our knowledge about the thermophysical parameters of sea ice. Moreover, the thermophysical parameters help us to interpret the data of remote studies of sea ice sheets in a deeper and more comprehensive way.

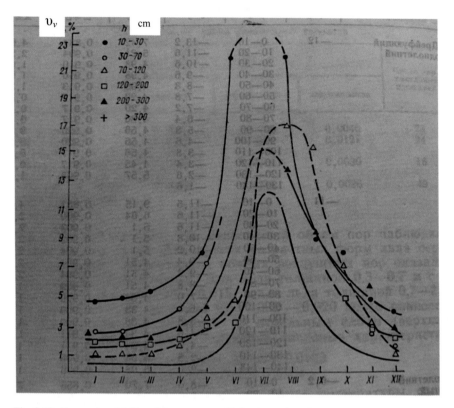

Fig. 3.10 Seasonal variability of the presence of brine in sea ice of various ages

3.4.1 Thermal Deformation of Ice

The measures of thermal deformation of ice are the coefficients of linear and volumetric expansion. They are measured as relative changes in length or volume, when the temperature changes by 1 °C. The laws of thermal expansion of sea ice differ from the general ideas about the expansion of other bodies [60]. The difference is that the change in volume with temperature changes consists of the usual compression of solid and liquid components of sea ice and an increase or decrease in volume due to the release or dissolution of fresh ice by brine. Depending on the prevailing process, either expansion or compression of the ice will occur (Table 3.9). The table shows, that up to a temperature of −7 °C, any decrease in temperature leads only to compression. And from a temperature of −17 °C and below, any decrease in temperature leads only to expansion.

Table 3.9 The coefficient of volumetric expansion of sea ice

Salinity of ice, %₀₀	Temperature of ice (°C)													
	-1.8	-4	-7	-8	-9	-10	-11	-12	-13	-14	-15	-16	-17	-18
1	-1327		78	99	113	124	134	140	145	150	154	151	160	163
2	-2823 (-2210)	(-412)	-13	29 (16)	36	80 (83)	98	111 (113)	122	131 (123)	140	145 (127)	151	156 (133)
3	-4320		-104	41	0	35	63	82	98	112	125	134	142	150
4	-3816 (-4589)	(-992)	-195	-110 (-137)	-57	-9 (-2)	27	53 (56)	74	98 (78)	93	122 (85)	133	144 (96)
5	-7318		-286	-181	-113	-54	8	23	51	73	96	110	123	137
6	-8808 (-6967)	(-1573)	-377	-261 (-290)	-170	-99 (-88)	44	6 (0)	37	54 (33)	32	98 (43)	114	131 (60)
7	-10.305		-468	-321	-226	-143	-19	-35	3	35	67	86	105	124
8	-11.301 (-9346)	(-2153)	-559	-392 (-443)	-282	-188 (-173)	-115	-64 (-57)	-2	16 (-13)	53	74 (2)	96	118 (23)
9	-13.297		-640	-462	-329	-232	-150	-33	-11	0	38	51	67	112
10	-14.793 (-11.725)	-2734	-741	-532 (-595)	-395	-277 (-259)	-186	-68 (-113)	-22	-23 (-59)	23	39 (-40)	78	105 (-13)
11	-16.289		-832	-602	-452	-321	-221	-151	-91	-41	9	230	69	99
12	-17.786		-923	-672	-508	-366	-257	-150	-115	-60	-6	56	60	93
13	-20.282		-1014	-742	-564	-421	-292	-210	-139	-70	-20	3	51	86
14	-20.778		-1105	-812	-621	-455	-328	-239	-162	-99	-35	-8	49	80
15	-22.764 (-17.672)	(-4285)	-1196	-882 (-978)	-667	-500 (-473)	-363	-368 (-254)	-130	-118 (-172)	-40	-10 (-145)	-33	78 (-104)

Note Malmgren's data is given in parentheses, and Saveliev's data is given without parentheses

Fig. 3.11 Change of thermal conductivity by sea ice thickness h is the thickness of the ice sheet; h1 is the thickness of the measurement layer; 1 is the area of thermal conductivity values for annual and long-term ice; 2 is the area of thermal conductivity values of sea ice by 1 m thick near Dixon Island

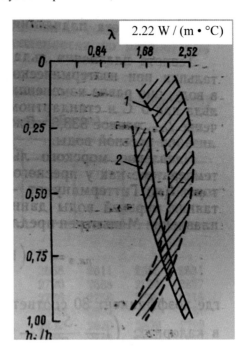

3.4.2 Thermal Conductivity and Temperature Conductivity

The coefficient of thermal conductivity of a solid body is defined as the ratio of thermal energy, that passes through a unit area per unit time, provided that the temperature gradient is perpendicular to the elementary site [61–64]. At a temperature, close to 0 °C, the thermal conductivity coefficient of freshwater ice is approximately 2.22 W/(m · °C). This value is approximately 4 times greater than the thermal conductivity coefficient of pure water at 0 °C.

Experimental data on the calculation of the thermal conductivity of sea ice are presented in Fig. 3.11. The thermal conductivity of sea ice decreases with an increase in its salinity and increases with an increase in the brine concentration, i.e. with a decrease in temperature. This is confirmed by the experiment.

The results of measurements of ice of different salinity and at different temperatures are presented in Table 3.10.

3.4.3 Heat of Melting and Specific Heat of Sublimation

The melting heat of ice L_{mel} is defined as the change in enthalpy during the isothermal transformation of a unit of mass of ice into water, and it is equal to the change in its internal energy. For pure ice at 0 °C and standard atmospheric pressure, an L_{mel}

Table 3.10 Dependence of the coefficient of temperature conductivity of ice on its salinity and temperature

Temperature (°C)	Salinity ($^0/_{00}$)				
	2	5	10	15	20
−5	4.76	4.2	3.79	2.28	1.40
−10	4.78	5.01	4.22	3.21	2.58
−15	5.05	5.05	4.09	3.68	3.45

Table 3.11 Effective melting heat of sea ice (кJ/кg)

Salinity of ice ($^0/_{00}$)	Temperature of ice (°C)							
	−2	−6	−10	−14	−18	−22	−26	−30
0	338	346	355	363	371	380	388	396
4	301	332	345	355	359	372	381	389
8	263	319	337	348	357	367	378	386
15	197	294	320	335	347	358	372	382
25	102	262	297	317	332	345	364	376
35	7	228	274	299	317	332	356	369

value equal to 333.5 kJ/cd was obtained. It coincides with the heat of crystallization of freshwater. The melting of sea ice does not occur at a certain temperature, as in fresh ice, but continuously, starting from a certain (according to Gitterman from − 36 °C) temperature to the melting temperature of seawater of a given salinity [65].

The results of calculations of the effective melting heat of sea ice are presented in Table 3.11. The effective melting heat decreases sharply with increasing salinity.

The latent heat of ice sublimation L_s is a change in enthalpy during the isothermal transformation of a single mass of ice into water vapor. At standard atmospheric pressure at the triple point of ice—liquid—vapor (273.16 K), the specific heat of ice sublimation is 2838 кJ/кg. The values of the specific heat of ice sublimation at different temperatures and salinities, according to the results of calculations, are given in Table 3.12.

3.4.4 Heat Capacity

The heat capacity is one of the main thermodynamic characteristics of ice. In practical calculations, the specific heat capacity of c_i is usually used, meaning by this a physical value, numerically equal to the amount of heat, that needs to be communicated to a unit of ice mass, in order to increase temperature by 1 °C [66].

According to experimental data, the heat capacity of hexagonal ice at the melting point at constant pressure is $c_p = 37.7$ J/ (mole · °C). This value is equal to about

Table 3.12 The specific heat of ice sublimation (кJ/кg)

Temperature (°C)	Salinity (⁰/₀₀)								
	0		5	0	15	20	25	30	35
	Water	Ice							
0	2101	2834							
−1	2504	2834	2751	2663	2574				
−2	2505	2835	2797	2750	2704	2658	2611	2565	2531
−3	2508	2835	2816	2788	2752	2720	2688	2656	2657
−5	2513	2835	2816	2817	2797	2776	2756	2736	2715
−8	2519	2835	2856	2842	2829	2816	2804	2793	2781
−10	2524	2836	2867	2868	2843	2832	2820	2809	2798
−15	2536	2837	2892	2883	2874	2865	2856	2846	2837
−20	2547	2837	2916	2908	2905	2902	2856	2879	2872
−25	2559	2837	2941	2938	2938	2941	2949	2964	2986
−30	2571	2837	2964	2960	2957	2958	2950	2947	2944

half of the heat capacity of water. The heat capacity of ice decreases with decreasing temperature, tending to zero at 0 K. The heat capacity of ice at a constant volume of c_v is 3% less than at c_p melting temperature. According to data of numerous experiments, the heat capacity of fresh ice at normal pressure and temperature T in °C can be determined, using the empirical formula [58, 67–69]:

$$c_i = (2.12 + 0.0078\,T)J/(g \cdot °C). \tag{3.5}$$

The calculated and experimental values of the effective heat capacity of sea ice as a multiphase system are given in Table 3.13. It follows from the given data that the heat capacity of sea ice at high temperature and high salinity can reach very large values. This is explained by the fact, that, when the temperature changes in salt cells, a significant amount of pure ice is formed or melted with the release or absorption of melting heat. Consequently, for sea ice, the concepts of heat capacity and heat of melting are inseparable from each other.

3.4.5 Volumetric Diffusion, Condensation, and Evaporation

Due to thermal fluctuations of molecules and the presence of defects in the ice lattice, water molecules can shift from equilibrium positions and diffuse. The simplest formula for the diffusion of atoms or molecules through a solid body is derived on the assumption, that the flow of molecules I is proportional to the gradient of their concentration [70]:

Table 3.13 Effective heat capacity of sea ice (кJ/кg·К)

Salinity of ice (⁰/₀₀)	Temperature of ice (°C)							
	−2	−6	−10	−14	−18	−22	−26	−30
0	2.09	2.05	2.05	2.01	1.97	1.93	1.93	1.88
2	11.3	3.10	2.14	2.22	2.09	2.05	2.05	1.97
4	20.4	4.14	2.72	2.43	2.26	2.13	2.22	2.01
6	29.6	5.19	3.10	2.64	2.39	2.22	2.39	2.09
8	38.7	6.24	3.43	2.81	2.51	2.34	2.51	2.14
10	47.9	7.28	3.77	3.01	2.64	2.43	2.68	2.22
15	70.8	9.88	4.65	3.52	2.97	2.64	3.06	2.34
20	93.7	12.5	5.53	4.02	3.31	2.89	3.43	2.51
25	117	15.1	6.40	4.56	3.64	3.14	3.85	2.68
30	139	17.7	7.28	5.07	3.98	3.39	4.23	2.85
35	162	20.3	8.08	5.57	4.31	3.60	4.60	3.01

$$I = -D_\upsilon \cdot \mathrm{grad}\, n = -D_\upsilon \cdot \nabla n \qquad (3.6)$$

where I is expressed in $\mathrm{M}^{-2} \cdot \mathrm{c}^{-1}$; n is the number of molecules in $1\,\mathrm{m}^3$; D_υ is a constant at a given temperature, equal, according to experimental data, to approximately 10^{-16} m^2/s. It is called the volume diffusion coefficient. The condensation coefficient α_c shows the probability, that a molecule, that has hit the surface is adsorbed, enters thermal equilibrium with the substance, connecting with its surface [71]:

$$\alpha_c = \frac{A_f \sqrt{2\pi\, mkT}}{P_s - Pm}$$

where A_f—full flow of matter to the surface; P_m—vapor pressure at temperature T; P_s—steam pressure in the environment ($P_s > P_m$; m—mass of a substance molecule; k—Boltzmann constant. The evaporation coefficient is determined similarly [72]

$$\alpha_e = \frac{A_f \sqrt{2\pi\, mkT}}{Pm - Ps}$$

where A_f—the total flow of matter from the surface, while $P_m > P_s$.

It has been experimentally established, that α_c for ice at a temperature, close to the temperature of the ice–liquid phase transition and α_w for liquid water have similar values. This indicates the existence of a «liquid film» on the ice surface at a temperature, close to the melting point. Note also that this circumstance certainly plays an important role, when considering the interaction of electromagnetic waves with the ice cover. The evaporation coefficient for ice α_e is estimated at 0.67.

3.5   Mechanical Properties

The mechanical properties of ice are characteristics of its behavior as a solid body under the action of internal and external forces. They are characterized by mechanical stress, deformation, work, and other physical parameters, which can be used to judge the ability of ice to its elastic, elastic-plastic, and viscous deformation or destruction [73–76]. The mechanical properties of ice include its elastic, plastic, and strength characteristics, such as, for example, the elastic limit, the strength limits for tensile, compression, shear or bending deformations, the modules of longitudinal and shear elasticity, the limit of long-term creep, as well as additional characteristics such as hardness, internal and external friction, adhesion and others.

The peculiarity of ice in comparison with other materials is that in natural conditions ice is at temperatures, close to the melting point, does not enter into chemical interaction with impurities and has a relatively large-crystalline structure. Studies show that ice has a low elastic limit and, even under low loads exhibits pronounced properties, that manifest themselves in the form of a decrease in strength over time, stress relaxation, and the development of creep deformations.

Ice in an elastic state, the relationship between mechanical stresses and deformations is described by the relations of the theory of elasticity; in the elastic–plastic state—by the equations of the theory of plasticity [70, 72, 77].

Experimental study of mechanical stresses is usually carried out by direct contact methods (for example, strain gauges). An indirect assessment of the stress state of the sea ice cover can be made, if necessary, by remote methods.

3.5.1   Elasticity and Internal Friction of Ice

The values, characterizing the elastic properties of ice, include its elastic modulus, Lame constants, and Poisson's coefficient. For small deformations, the relationship between the stress components σ_x, σ_y, ..., τ_{xy} and the strain components ε_{xx}, ε_{yy}, ..., ε_{xy}, at some point, is generally represented by six linear relations of the type [78]

$$\sigma_x = c_{11}\varepsilon_{xx} + c_{12}\varepsilon_{yy} + c_{13}\varepsilon_{zz} + c_{14}\varepsilon_{yz} + c_{15}\varepsilon_{zx} + c_{16}\varepsilon_{xy}$$

where σ_x—the normal voltage, parallel to the axis on the platform with the normal x, etc.; τ_{xy}—tangential stress, parallel to the y axis along the platform with the z normal, etc.; ε_{yz} is a relative shift, i.e. a change in the yz plane of the right angle of an elementary parallelepiped, etc.

The coefficients c_{11}, c_{12}, ..., c_{66} are called elastic modules and have the dimension of stress, i.e. the unit of force, assigned to the unit of area.

For hexagonal crystals like single crystals of ice, there are five independent elastic modules, that are not equal to zero and are also called stiffness constants (c_{11}, c_{12}, c_{13}, c_{33}, c_{44}).

When considering the basic elementary types of the stressed state of ice (unilateral normal stress, pure shear, and all-round normal stress), the physical meaning of elastic modulus is revealed. The relationship between the stress and the corresponding deformation in each case is determined by the simplest formula: the stress is equal to the product of the corresponding deformation by the elastic modulus.

Elastic modulus determines the macroscopic properties of pure ice and its constants, which are found experimentally or by calculation, based on the microscopic theory of interatomic bonding forces. The theoretical constants of elasticity of a single crystal of ice were calculated from the data on the parameters of the crystal lattice of ice, based on the dynamic Born model. The stiffness constants c_{ij}, which are coefficients in the generalized expression of the stress tensor through the deformation tensor, expressed in 10^{-9} Pa, turned out to be equal [79]:

$$c_{11} = 15, 2; c_{12} = 8, 0; c_{13} = 7, 0; c_{33} = 16, 2; c_{44} = 3, 2.$$

Five independent elastic constants were also calculated from the values of the propagation velocities of longitudinal and transverse waves, respectively oriented with respect to the optical axis of the ice crystal, from which the samples for ultrasonic sensing were made. The experimental values are in good agreement with the above-calculated data.

The elastic properties of a sea body are different from the elastic properties of a single crystal. Sea ice is a relatively fine-grained physical body, which in the freezing plane can most often be considered statistically isotropic due to the chaotic azimuthal orientation of the optical axes of individual crystals. Temperature gradients and inhomogeneities of the structure along the thickness of the ice cover can lead not only to vertical, but also to horizontal anisotropy of ice, which sometimes manifests itself over significant areas.

Before proceeding to the numerical properties of sea ice, let us briefly dwell on the rather diverse terminology, encountered in describing the elasticity of ice. In addition to the basic «elastic» term «Young's modulus», which is the modulus of longitudinal elasticity, we present the following concepts, showing a particular feature of this property and the degree of approximation to its definition [80].

The true or instantaneous Young's modulus is a value, corresponding to instantaneous elastic deformation, for example, in an idealized uniaxial compression experiment, in which it is assumed, that stress is applied suddenly at some point in time and an infinite rate of stress change and an infinite deformation rate occur. In typical uniaxial compression experiments, the elastic deformation has a value of 10^{-4} at a stress of about 10^5 Pa. If, on the «stress—deformation» diagram, at the initial moment of time, corresponding to the origin of coordinates, draw a tangent to the curve $\sigma = f(\varepsilon)$, then its slope $(d\sigma/d\varepsilon)_{\varepsilon = 0}$ is *the true* Young's modulus, which is also called with *the initial tangent modulus*.

The average value of the derivative $d\sigma/d\varepsilon$ is also called *the secant* module. At high deformation rate or low temperature, this average value is large, at low deformation rate or high temperature, the curvature of the diagram $\sigma = f(\varepsilon)$ becomes very significant and the secant modulus is noticeably smaller than the true modulus $(\Delta\sigma/\Delta\varepsilon)_{\varepsilon=0} < (d\sigma/d\varepsilon)_{\varepsilon=0}$.

Acoustic (ultrasonic, seismoacoustic) is usually called Young's modulus, determined by the measured values of the density ρ and the velocity of propagation of longitudinal C_l and transverse C_t waves in a sample on a small base (units— tens of centimeters) in the case of using the pulsed ultrasonic method or in the ice cover, when the measuring base is tens to hundreds of meters, in the case of seismic method using [72, 74]. Young's modulus, determined from measurements on large bases, is sometimes called *an average integral*, or *an effective dynamic module*.

The dynamic modulus of elasticity, which gives a connection between the final changes σ and ε, which are carried out so quickly, that no relaxation has time to occur, is called the *nonrelaxing modulus* $E_n = \Delta\sigma/\Delta\varepsilon$. *The relaxed modulus of elasticity* is a modulus for those conditions, when the relaxation process has already ended [81]:

$$E_n = E_p \frac{\tau_\delta}{\tau_\varepsilon}, \qquad (3.7)$$

where τ_ε—stress relaxation time at constant deformation; τ_σ—the time of retardation, i.e. the value, characterizing the rate of increase in deformation at constant stress.

The theory of mechanical relaxation of ice is based on statistical consideration of the structure and equilibrium state of the electrostatic interaction of hydrogen bridges of the ice lattice. It is known that oxygen atoms in an ice crystal occupy a hexagonal spatial lattice. The arrangement of hydrogen atoms inside this basic lattice is uncertain, and they are characterized by frequent transitions. All possible configurations should occur with approximately the same probability. Based on the statistical structural model, the existence of a maximum of the logarithmic decrement (attenuation coefficient) of oscillations or the tangent of the angle of mechanical losses is explained by the fact that sudden deformation of the crystal causes additional energy differences between different complexes of hydrogen atoms; the dynamic equilibrium between these complexes is disturbed. After the relaxation time has elapsed, a new equilibrium occurs, at which, on average, energetic, favorable configurations of complexes of hydrogen atoms occur more often [82]. Based on the above considerations, the diffusion of atoms due to defects in the crystal lattice can be attributed to one of the possible mechanisms of dissipation, caused either by damage to the lattice, or by chemical impurities or inclusions. The dissipation energy is 356 kJ/mol. This value is equal to the activation energy.

The characteristics of the internal friction of freshwater polycrystalline ice have been studied in some detail in laboratory conditions on small samples. It is established, that there can be three types of mechanical attenuation: caused by processes at grain boundaries, proton motion, and chemical impurities. The resulting activation energy due to the movement of protons is approximately equal to 25.1 kJ/mol. Since the deformation of ice changes the parameters of the crystal lattice, we can expect a change in the spectrum of its own radio emission [83]. It is still difficult to answer, how significant these changes are and whether they can be used to improve the remote assessment of the stressed state of the sea ice cover. Apparently, the reason for this is not only the weak knowledge of the processes of mechanical and dielectric relaxation of sea ice and the expected effect of changes in the spectrum, which is small in terms of energy, but also the conclusion about strong shielding effect of temperature fluctuations of radio-brightness radiation of ice.

For periodically varying voltages $\sigma(t) = \sigma_0 e^{i\omega t}$ and deformations $\varepsilon(t) = \varepsilon_0 e^{i\omega t}$, the relationship between their amplitudes can be obtained in the form $\sigma_0 = E^* \varepsilon_0$, where E^* is called *complex module* [80]:

$$E^*(\omega) = E_1(\omega) + iE_2(\omega) = \frac{1 + i\omega\tau_\delta}{1 + i\omega\tau_\varepsilon} E. \tag{3.8}$$

Tangent of angle φ, by which the deformation is delayed relative to the stress and which is a measure of internal friction, will be equal to the ratio of the imaginary part $E^*(\omega)$ to the actual $1 + \omega^2 \tau_\varepsilon \tau_\delta$.

$$\text{tg}\,\varphi = \frac{\omega(\tau_\delta - \tau_\varepsilon)}{1 + \omega^2 \tau_\delta \tau_\varepsilon}. \tag{3.9}$$

Usually, τ_σ and τ_ε differ by several percent and therefore either do not distinguish between them, or use their geometric mean. If $\tau_m = \sqrt{\tau_\delta \tau_\varepsilon}$, and $E = \sqrt{E_n E_p}$ the tangent of the angle of mechanical losses, where τ_m—mechanical relaxation time, which depends on temperature T and which, through the activation energy of mechanical relaxation W and the Boltzmann constant k, can be estimated, using the formula $\tau_m = \tau_0 e^{W/kT}$. The value $\Delta_E = (E_n - E_p)E$ characterizes the degree of relaxation and is called *Young's modulus defect* (sometimes the denominator is put in place of the E value E_p). The degree of relaxation for the shear modulus can be obtained from the expression [80]

$$\Delta_G = \frac{1}{2(1 + v)} \Delta_E, \tag{3.10}$$

from which it can be seen that the value Δ_G about 10–15% more Δ_E. *The dynamic modulus of elasticity $E(\omega)$*, determined by the ratio of stress to the part of deformation, that is in phase with stress, corresponds to the expression [84]

$$E(\omega) = E_n - \frac{E_n - E_p}{1 + (\omega\tau_m)^2} \tag{3.11}$$

where from

$$E(\omega) = \begin{cases} E_p & when\ \omega\tau_m \ll 1 \\ E_n & when\ \omega\tau_m \gg 1. \end{cases} \tag{3.12}$$

There are also *isothermal and adiabatic elastic modulus*. During isothermal deformation, the body temperature does not change, and the elastic modules, corresponding to this case, are called isothermal. In the case of adiabatic deformations, the modules are determined with sufficient accuracy by the expression [84]

$$E_{ad} = E + E^2 \alpha_v \frac{T}{c_p} \qquad (3.13)$$

where α_v—coefficient of volumetric thermal expansion; c_p—specific heat capacity at constant pressure; T—the temperature of the deformed body. For ice, the difference between adiabatic and isothermal modules is small. For example, in polycrystalline freshwater ice at a temperature of about $-10\ °C$, their calculated values differ by about 0.1%.

Modulus of deformation. This term was proposed, in order to more clearly distinguish the true modulus of elasticity of ice from its quasi-property, which manifests itself during relatively long deformation processes in static experiments on bending, stretching, or compressing samples. The apparent elastic modulus, obtained in these cases with the help formulas of resistance of materials, linking the deformation of ice with the load, that caused it, are called, respectively, deformation modules during bending, stretching, or compression. These values depend significantly on many factors, including the method of conducting experiments, the size of samples, and the method of registering deformation.

Due to the difference in the properties of ice for compression and tension, the modulus of deformation during bending is a reduced modulus. In the resistance of materials, it is shown, that it is expressed in terms of the elastic modulus of compression $E1$ and tensile $E2$ by the following formula [84]:

$$E_{red} = \frac{4E_1 E_2}{\sqrt{E_1} + \sqrt{E_2}} \qquad (3.14)$$

where $E_1 = E_2$, $E_{red} = E_1 = E_2$. Thus, the reduced modulus of elasticity is such a modulus, which is obtained, if we assume, that the material works equally for both compression and tension.

The shear modulus corresponds to the stress state of pure shear, in which only tangential stresses τ_t act on two mutually orthogonal platforms in the vicinity of a certain point of the body. It is equal to the ratio of the tangential stress τ_t to the shear angle γ, which determines the distortion of the right angle between the planes, along which the tangential stresses τ_t acts, and determines the ability of the material to resist shape change, while saving its volume [85]. Sometimes, the term «shear modulus» is also used with «distortion modulus», «transverse modulus »and «chipping stress modulus». Other terminological features noted definition of the term «Young's modulus» also apply to the shear modulus. The values of the shear modulus of ice are 2.6–2.7 times less than Young's modulus.

The modulus of *volume compression* (or *volume modulus of elasticity*) corresponds to the all-round equal normal stress σ_n, which occurs at hydrostatic pressure.

It is equal to the ratio of the normal voltage to the relative volumetric compression, caused by this voltage [81]:

$$K = \frac{\sigma_n}{\Delta}$$

where $\Delta = \varepsilon_{xx} + \varepsilon_{yy} + \varepsilon_{zz}-$ relative change of volume. K characterizes the ability of ice to resist changes in its volume, not accompanied by changes in shape.

The values of the volumetric modulus of elasticity are usually several percent higher than the values of Young's modulus. However, for some porous layers of long-term sea ice, they may be a few percent smaller.

To assess the seasonal variability of Young's modulus, experimental and calculated data are used, that characterize the particular dependences of this elasticity characteristic on both internal parameters and properties of ice (average crystal size and their distribution, isotropy and anisotropy of the granular structure and crystallographic orientation, the presence or absence of impurities and inclusions, such as air bubbles, salt solution, etc.), and from external conditions (the values of the load, applied to the ice, which determines its stress state; the temperature of the ice; the time of application of the load or the frequency of changes in mechanical stresses in the ice under cyclic loads; the deformation rate). Moreover, in some cases, a large-scale effect also affects these dependencies [85].

Of the variety of partial dependencies of Young's module, available in publications, those, that were useful for obtaining a graph of seasonal variability of the dynamic Young's module of sea ice, are taken into account. For freshwater polycrystalline ice, various authors have obtained temperature dependences almost identical in slope, differing in absolute value [86–88]. Thus, the dynamic modules, obtained by Kuroiva and Jamaica, and the values of the quasi-dynamic module, obtained by Gold by the static method, close to them, are on average 20–30% smaller than the modules, determined respectively by Berdennikov formulas, if the latter is taken as the initial ones, when compared.

The calculated dependences of Young's modulus on the average size of ice grains, on the stress in the ice during measurement, on the time (frequency) of load application and on temperature, obtained by Synch, and their comparable analysis with experimental data allow us to understand the large spread of published data on Young's modulus of freshwater ice, especially in cases, where there is no sufficient information about the properties of the ice itself and the conditions of the experiments.

However, it should be noted, that if, according to numerous measurements, the elastic modulus of polycrystalline ice, determined by static methods (with a short-term application of a full load cycle for 5–10 s), are in the range of 0.3–11.0 GPa, i.e. differ by 30 times, then the elastic constants, determined by the pulsed ultrasonic method are more stable characteristics: for example, at a temperature of $-5\ °C$, 11 different authors have published the values of Young's modules of freshwater ice, differing by an average of 10%.

Compared with freshwater ice, sea ice has a much greater dependence of its physical properties on temperature due to the content of an aqueous solution of salts (brine) in it [89]. And it is brine inclusions (their geometric shape, dimensions, and distribution in ice), that lead to the fact that the mechanical behavior of the composite material «pure ice plus salt inclusions» is determined by effective elastic constants and effective strength characteristics, which can be set as functions of the elastic constants of pure ice at a certain temperature and concentration (volumetric porosity) salt inclusions. The generalized results of numerous studies of the elasticity properties of long-term sea ice, performed by the seismoacoustic method by various authors, are presented in Table 3.14.

The obtained laws of the distribution of elasticity characteristics over the thickness of long-term ice at different times of the year are also valid for annual ice sheets. However, seasonal changes in these parameters should be more pronounced in them due to their increased salinity [89]. Data analysis shows, that the values of elastic modulus, weighted average by the thickness of the layer of 30–150 cm, are 20–25% lower for annual ice, compared to long-term ice at comparable temperatures.

Currently, there are a number of empirical formulas allowing us to calculate Young's modulus, taking into account the volumetric porosity of ice.

In particular, for $v_a \leq 0{,}15$ it is found that Young's modulus varies linearly, depending on v_a. There are no experimental data for cases of increased porosity. However, they can be obtained by extrapolation. This happens as follows [90]:

Table 3.14 Generalized elasticity characteristics of 300 cm thick long-term sea ice

Layer of ice (cm)	Salinity ($^0/_{00}$)	Temperature (°C)	Density (kg/m^3)	Velocity of longitudinal waves (m/s)	Young's modulus 10^3 MPa	Coefficient of Poisson
Summer-autumn period						
50–300	2.6 ± 0.8		910–925		6.6–7.5	0.32–0.35
0–50		−1, …, −15		3162 ± 261		
		−1, …, −9				
50–200		−1, …, −9		3387 ± 162		
		−2, …, −5				
200–300		−2, …, −5		3206 ± 126		
		−2, …, −4				
Winter-spring period						
50–300	2.1 ± 0.6		910–925		8.0–8.4	0.32–0.35
0–50		−15, …, −38		3650 ± 240		
		−9, …, −29				
50–200		−9, …, −29		3768 ± 96		
		−5, …, −16				
200–300		−5, …, −16		3533 ± 300		
		−4, …, −7				

$$E/E_0 = \begin{cases} 1 - v_a, 0 \le v_a \le 0.12; \\ 47.168(0.15 - v_a)^3 - 45.97(0.15 - v_a)^2 + 0.5(0.15 - v_a + 0.25), 0.15 \le v_a \le 0.4; \\ 0.06(1 - v_a), 0.4 \le v_a \le 1; \end{cases} \quad (3.15)$$

$E_0 = 9.21 \, (1 - 0.001467 \, T)$, GPa—Young's modulus of pure ice as a function of temperature T.

Other empirical formulas are also known, but in our calculations we will use the formula, experimentally verified by Frederking and Hasler to estimate the seasonal variability of the Young modules of sea ice (Fig. 3.12):

$$E/E_0 = \left(1 - \sqrt{v_a}\right)^4, \quad (3.16)$$

for which $E_0 = 9.21 \, (1 - 0.001467 \, T)$, and dependence, much earlier than the above, proposed by V.P. Berdennikov:

$$E = E_0 \left(1 - 3\frac{s_i}{s_b}\right), \quad (3.17)$$

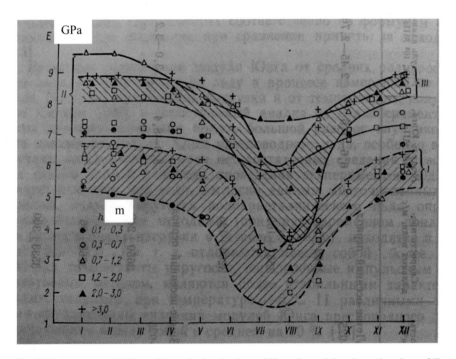

Fig. 3.12 Seasonal variability of the calculated values of Young's modules. *I*—static values of *E*, according to L. Gold's formula; *II*—dynamic values of *E* by averaged values of longitudinal wave velocities; *III*—dynamic values of *E*, according to V.P. Berdennikov's empirical formulas

in which $E_0 = 87.6 - 0.2(T_i - 0.0017T^2_i) \times 10^8$ Pa; S_i—salinity of ice, ‰; S_b—brine salinity at a given ice temperature T_i.

These two formulas give very close calculated values of dynamic Young's modules of sea ice, and the calculation of Young's modules, based on the values of density and velocity of longitudinal ultrasonic waves, using the formula [90]

$$E = \rho c_1 \frac{2(1 + v)(1 - 2v)}{1 - v},\tag{3.18}$$

which can be considered independent in comparison with the calculations, according to the two formulas above. Figure 3.12 also shows the values of static Young modules of sea ice, calculated by the formula $E = E_0 (1 - v_a)^4$, $E_0 = 5.69 - 0.0648T_i$ GPa. Analysis of the graphs shows that less salty thick ice has greater elasticity, and the difference in the values of Young's modulus for ice gradations $h = 10$–30 cm and 200–300 cm persist throughout all months of the year and is in the range of 15–100% or more.

The periods of the greatest variability of Young's modulus is the period May–July, when the modulus decreases 2–3 times during the transition from winter to summer (see, for example, the family of modules I in Fig. 3.12, where Young's modulus of ice with a thickness of 120–200 cm was estimated in May with a value of $E = 5.5$ GPa, and in July the same ice corresponds to the calculated value of $E = 2$ GPa) and the period August–November, when the ice modulus increases 2–3 times as it decompresses and cools (see, for example, the values of ice modules at $h = 120$–200 cm in August and November, equal respectively to about 2 and 6 GPa).

The general variability of the calculated values of Young's dynamic modulus for all ice gradations and throughout the year (approximately 2–9.5 GPa) fits well into the range of known experimental data (1.7–9.1 GPa). However, Young's modulus varies significantly for the average layers of long-term Arctic ice (values $E = 6.6$–7.5 GPa for the summer–autumn period and values $E = 8.0$–8.4 GPa for the winter–spring period in Table 3.14). The Poisson's coefficient for the studied layers of annual and long-term ice sheets (see Table 3.14) is estimated and the values are mainly in the range of 0.32—0.35. There are no measurements that can be used to trace the seasonal variability of the Poisson's coefficient for various ice. You can make a recommendation to use in the calculations the formula

$$v = 0.333 + 0.06105 \exp(T_i/5.48),\tag{3.19}$$

which gives the following extreme values near the melting point and at very low temperatures (theoretically—∞) $0.333 \leq v \leq 0.394$.

3.5.2 Plasticity of Ice

The property of ice to irreversibly deform under the action of external forces or internal stresses is called plasticity [91]. At the same time, some elements of the volume of the deformable body experiences displacement, and the measure of irreversible changes in the mutual disposition of the particles are the components of the plastic distortion tensor, the symmetric part of which is the plastic deformation tensor ε^p, whose components (provided the elastic deformation is small, compared to plastic $\varepsilon = \varepsilon^E + \varepsilon^p \approx \varepsilon^p$) are not uniquely related to the components of the stress tensor σ_s.

Since ε^p is not a function of the state, there are no ice characteristics, connecting ε^p and σ_s, which could be considered its constants (like the elasticity constants, connecting elastic deformation of ε^p and σ_s). The rate of plastic deformation ε^p depends on the instantaneous values of σ_s, temperature T, and ice structure. When there is a linear coupling of ε^p (σ_s) in the presence of nonequilibrium defects in the ice structure (for example, dislocations), plastic deformation is called *a quasi-viscous flow of ice.*

When we say «ice flows», then we usually mean its viscous flow by analogy with a Newtonian viscous fluid, for which the flow velocity linearly depends (through the flow coefficient A) on pressure, i.e. in relation to ice, a ratio of the type [92, 93] must be observed

$$\varepsilon^p = A\sigma_s.$$

Moreover, A is a function of temperature, grain size, and parameters of the dislocation process of atom diffusion in the ice crystal lattice. In terms of dimension, the coefficient A is equal to the inverse value of the *viscosity coefficient* ($A - 1/\eta$). The experimentally observed instability of the viscosity coefficient is associated with the increase in stresses over time of the properties of ice from the properties of viscosity. The term «viscous flow of ice» (or, more correctly, «quasi-viscous»), apparently, can be used for ice at «warm» temperatures, when the so-called non-threshold deformation prevails, i.e. the deformation that occurs in it with small applied loads. It is known that the creep process (namely its viscous mechanism) is facilitated in sea ice due to the presence of a liquid phase in it—a salt solution.

Thus, the viscosity of ice characterizes its resistance to the development of residual deformation in it under the influence of external forces. Viscous flow is observed at stress, that is less than the yield strength and is characterized by the fact, that the deformation rate decreases with decreasing stress and vanishes, when it is removed. Quantitatively, the viscosity coefficient (or internal friction coefficient) can be determined through the tangential force F, which must be applied to the unit area S of the shear layer, in order to maintain the laminar flow in this layer at a constant rate of relative shear ε^p [94]:

$$\eta = \frac{F}{S}\frac{1}{\varepsilon_p} = \frac{\tau_c}{\varepsilon_p}, \qquad (3.20)$$

where τ_c—stress of shear.

The viscosity coefficient is used for torsion, shear, stretching, compression, and bending deformations. One of the most common methods is the method of bending a free beam, lying on two supports, in which the viscosity coefficient is calculated by the formula [94]:

$$\eta = \frac{Fl^3}{12bh^3v}. \qquad (3.21)$$

Here v—steady-state rate of plastic deformation.

When stretching (compressing) a rod of length l and cross section S

$$\eta = \frac{Fl}{Sv}. \qquad (3.22)$$

The coefficient of dynamic viscosity can be determined by the decrement of mechanical vibrations δ of the sample at some frequency f and the known modulus of elasticity E, if the known ratio is used for these purposes [94]:

$$\eta = \frac{\delta E}{2\pi^3 f}. \qquad (3.23)$$

The experimental values of the viscosity coefficient obtained by static methods are so contradictory (from 10^9 to 10^{15} Pa $\cdot$ s) that it is difficult to establish any law of its change. Therefore, the viscosity coefficient of ice is practically a conditional value that characterizes the creep rate under given deformation conditions and at a given time. According to a number of authors, the viscosity of ice does not satisfy Newton's law due to the absence of a linear relationship between stress and deformation rate. The viscosity of the plastic flow of ice is 5–6 orders of magnitude greater than the viscosity of the same ice, calculated from the measured parameters of elastic vibrations.

The continuous change in time of plastic deformation of an ice body under constant stress is called *ice creep* [93, 95]. The physical mechanism of creep of a single crystal of ice is caused by the sliding of molecular layers of ice, parallel to the basic planes and located near them. Moreover, due to the movement and multiplication of dislocations (defects in the crystal structure of ice), this sliding is carried out at stresses of shear significantly lower than expected ones (theoretical). The change in the shape of the deformable body and the numerical value of the creep deformation of single crystals significantly depends on the direction of the applied load, relative to its C-axis. According to experimental data, the limiting stresses, that caused creep in the ice in directions, parallel to the basic planes, were 0.4 MPa, and in the perpendicular direction 9 MPa, i.e. more than 20 times.

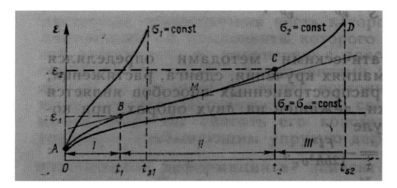

Fig. 3.13 Curves of creep for polycrystalline ice at three different stress levels: $\delta_1 = $ const $> \delta_2 = $ const $> \delta_3$. Stages of creep: *I*—primary, *II*—secondary, *III*—tertiary

In polycrystalline ice, due to the possibility of relative sliding of ice crystals and their spatial orientation, as well as due to the fact, that in a deformed array, the stress state is almost always inhomogeneous and, consequently, stresses of shear act in different directions, relative to the axes of crystals, the anisotropy of creep properties is less pronounced than in a single crystal of ice. Polycrystalline ice often behaves like an isotropic solid body.

The graph of the change in the deformation of ice ε as a function of time at constant stress is called curve of its creep. The stress-deformed state of ice is usually described by a family of curves by creep (Fig. 3.13), the nature of the change in time which depends on a number of parameters, including stress, temperature, structure, direction of load application, etc.

Polycrystalline ice as a plastic solid body is characterized by the following deformation patterns. During compression, for example, after elastic deformation, which occurs almost instantly (segment *OA* in Fig. 3.13), creep begins, which can be divided into three sections: *AB*—the stage of *unsteady creep* with a decaying deformation rate, this stage is also called *decelerating*, or *primary creep*; *BC* is a stage of *steady-state creep* with an almost constant rate of deformation, which is also called stationary mode, or *secondary creep*, which can last for a long time, if the stress and temperature of the ice are low enough, for this stage as a slow transition process is characterized by the presence of an inflection point M, the tangent in which gives the value of the speed of secondary creep; *CD*—*tertiary creep, accelerated creep*, or progressive flow, characterized by accelerated deformation, leading to the destruction of ice. The transition to tertiary creep is achieved as a result of ordering the orientation of ice crystals during recrystallization.

At stresses, not exceeding a certain critical value σ_∞, called the *limit of long-term creep*, the shear deformation of ice will increase continuously at a constant rate [96]. The value of σ_∞ is characterized by the fact, that at $\sigma_s > \sigma_\infty$ the ice flow is accompanied by the formation of microcracks, leading to volumetric expansion and a decrease in its density.

As the stress increases, the area decreases until it disappears, and when the stress increases above the ice strength limit, its brittle destruction occurs. The parameters of ice are usually determined from experimental curves of creep, with the help of which it is possible to compile a phenomenological equation of its state for various conditions of deformation.

Using a family of curves by creep, it is possible to construct a dependence of the type $\sigma_s = f(t_s)$, which is a convenient characteristic of the «durability» (indestructibility) of ice at a given load.

Figure 3.14 presents experimental data on the plastic behavior of sea ice.

It has been theoretically and experimentally established, that ice creep generally obeys Glen's power law

$$\varepsilon^p = k_e \sigma_{cr}^n \tag{3.24}$$

where k_e—empirical constant, n—the indicator of degree, varying in the range of 1.85–4.16 with a voltage change from 0.1 to 1.5 MPa.

Since the deformation and destruction of ice is considered as a single thermal fluctuation process, the deformation rate can also be expressed in terms of the energy of «crawling» dislocations W_{cr}, ice temperature T, and Boltzmann constant k [97]:

$$\varepsilon^p = e^{-(W_{cr}/kT)}. \tag{3.25}$$

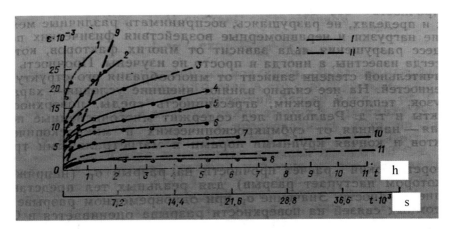

Fig. 3.14 Curves of creep for sea ice, I—according to pressiometric tests, performed in the ice sheet layer with a salinity of 4,5 °/₀₀ at a temperature of −2 °C for various voltages: 1—1.58 MPa, 2—1.38 MPa, 3—1.18 MPa, 4—0.98 MPa, 5—0.78 MPa, 6—0.58 MPa, 7—0.38 MPa, 8—0.22 MPa, II—according to tests for uniaxial compression of samples at a temperature of −27°C for 9—0.45 MPa, 10—1.35 MPa, 11—2.25 MPa

3.5.3 Ice Strength. General Provisions, Terminology, Definitions, Brief Information About Experimental Studies

Strength is a property of materials under certain conditions and limits, without breaking down, to perceive various mechanical loads and uneven effects of physical fields. The process of ice destruction depends on many factors, that are not always known, and sometimes simply not studied. The strength of ice largely depends on the variety of its structural features. It is strongly influenced by external conditions—the nature of loads, thermal regime, aggressiveness of the environment, surface effects, etc. Real ice contains numerous damages—ranging from submicroscopic and microscopic defects to large pores and main cracks.

The theoretical calculation of the strength for fracture σ_f (the stress at which the fracture occurs) for real bodies presents great difficulties. The value of σ_f with simultaneous fracture of all interatomic bonds on the fracture surface is estimated at 0.1 E, where E is Young's modulus. Usually, the experimental strength values are several orders of magnitude lower than the theoretical ones. The reason for the low strength of ordinary bodies is an uneven distribution of internal stresses, interatomic bonds are loaded differently, and there are weak points in the atomic structure of bodies. When the external and internal stresses of the same name are added together, local over voltages also occur, which can reach the values of theoretical strength, leading to the fracture of interatomic bonds. In weak places of the structure, under the influence of large local stresses, the fracture of interatomic bonds occurs very easily—this is how breaks in the continuity of the body arise. The growth and fusion of fractures of continuity forms a macroscopic crack, the development of which leads to the destruction of the body. Theoretical strength is also called *ideal strength*, the density of cohesion forces (i.e. the forces of molecular interaction of parts of the same body), or simply *cohesion*, which can be characterized by the heat (work) of evaporation [98].

According to *the mechanical concept* of strength, the fractures into parts is considered as a result of *the loss of stability of a solid body*, located in the field of external and internal stresses. It is believed that for each material there is a certain *threshold voltage*, after which the body loses stability and breaks. Below the threshold, the body is stable and can maintain integrity under load indefinitely. This threshold stress is taken as a measure of the strength of the body. In *the kinetic concept*, the main attention is focused on the process of destruction development. The fracture of the body is considered as the final stage of the gradual development and accumulation of submicroscopic destruction. This process develops in a tense body under the influence of thermal fluctuations. The concept of *durability under load*, i.e. about the time t_s, required for the development of the process from the moment of loading the body to its fracture, is introduced. The durability t_s of a body (or the time of its destruction) under a tensile load, the breaking stress σ_s, and the absolute temperature T are related by the ratio [99]

$$t_s = t_0 \exp\left(\frac{u_0 - \gamma \delta_s}{kT}\right) \tag{3.26}$$

in which t_0, u_0, γ are constant values, determined by the physicochemical nature of the solid body and its structure; k is the Boltzmann constant, and the energy barrier $u = u_0 - \gamma \sigma_s$ is close in value to the binding energy of atoms in the crystal (sublimation energy). In relation to freshwater ice, the initial energy barrier, or activation energy, $u_0 = 0.6$ kJ/kg, time constant $t_0 = 4.6 \times 10^{-17}$ s, coefficient, determining the degree of reduction of u_0 under the action of breaking voltage, $\gamma = 1.4 \times 10^{-26}$ m^3/mol. The law of «Zhurkov» denies the concept of strength limit. The question of what kind of load a body can withstand, i.e. what is its strength, without specifying the time, during which it should remain unbroken, has no unambiguous answer. This indicates, that the terms «strength limit», «limit breaking stress» are conditional. They lose their meaning, when judging the physical nature of the strength of solid bodies, but they are quite convenient for practice.

In the scientific and technical literature on ice, in which its strength properties are discussed, the following terms can be found, related to the category of its strength characteristics: *the strength limits (temporary resistances) of ice for tensile (breaking), bending, compression* (respectively: under uniaxial load or under complex types of loading) and *on shift (by slice)*. These characteristics are usually obtained at time intervals of loading samples, not exceeding several seconds from the beginning of loading to the moment of their destruction. Therefore, they are also called *short-term* (or *conditionally instantaneous*) ice *strength*, respectively, for fracture, bending, compression and shear.

The criteria for assessing the strength of ice as a material, if it is necessary to show its ability to resist for a long or some finite time, are [99]:

the limit of long-term strength (or *limit of creep*) σ_s is a conditional stress, which after some time t_s (called in this case *the time of fracture* at a given stress) «leads» the deformable ice to the final stage of tertiary creep, when its deformation rate (respectively, and deformation) tends to infinity ($\varepsilon_p \to \infty$ and $\varepsilon_p \dot{} \to \infty$), and the limit of *long-term creep* σ_∞, corresponding to the stress, at which the deformation practically does not change ($\varepsilon_p \dot{} \to 0$).

As follows from Fig. 3.15, at $\sigma_s > \sigma_{s12}$ (these are the stresses σ_{si}), the destruction of ice in accordance with the above occurs after times, equal to $t_{s1} < t_{s2}$. At a voltage of $\sigma_s < \sigma_{s1}$, the ice does not collapse. And at $\sigma_s > \sigma_{s1}$ ($t \ll t_1$), the flow platform on the stress–strain diagram does not have time to form, there is no shear deformation, corresponding to steady creep, progressive creep occurs quickly, which is the cause of *brittle ice destruction*. In this case, the term ice strength limit σ_a is applicable, which is sometimes also called *the brittle strength of ice*.

From the family of curves by creep, it is possible to obtain the so-called long-term strength curve, i.e. the $\sigma_s = f(t_s)$, which is a convenient characteristic of the «durability» (indestructibility) of ice at a given load (Fig. 3.15).

For some types of freshwater and sea ice, the following dependencies between t_s and σ_s are known [100]:

Fig. 3.15 Curves of long-term ice strength at temperature $T_1 < T_2$

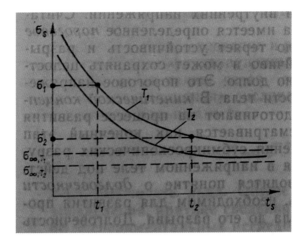

$$t_s = \frac{A^{\wedge}\left(\frac{1}{\lambda p}\right)}{(\sigma s - \sigma \infty)^{\wedge}\left(\frac{n}{\lambda p}\right)}; \quad \sigma_s = \sigma_\infty + \frac{A^{\wedge}\left(\frac{1}{n}\right)}{ts^{\wedge}\left(\frac{\lambda p}{n}\right)}, \quad (3.27)$$

in which A, n, λ_p, σ_∞—parameters are obtained from experiments on uniaxial compression of ice samples or, when ice is compressed directly in the array.

The experimental data, presented in Figs. 3.16 and 3.17, give an idea of the range of variability of strength characteristics by the example of its resistance to compression and bending moment. The limit of ice strength for bending $\sigma_{bend.}$ is the most important strength characteristic, which is widely used both in assessing the bearing capacity of the ice sheet and in calculating the forces of the impact of ice on various hydraulic structures. There are data on the seasonal variability of the $\sigma_{bend.}$, showing the regional features of the formation and existence of the ice cover. The fact is that differences in hydrometeorological conditions in vast sea areas cause the formation and development of an ice cover, that is heterogeneous not only in age, morphological characteristics, but also in the structure and physical properties of ice.

This is especially evident in the shelf zone of the Arctic seas, where the influence of freshwater runoff of Siberian rivers, shallow depths and sometimes complex bottom relief, stationary solder wormwoods is great. Almost all the main types of ice, that are available in the structural and genetic classification of ice in natural reservoirs, are found here [93]. The presence of sea, desalinated, and freshwaters not only determines the formation of ice, which differ in their structure and physical properties, but also contributes to the contact supercooling of water masses with abundant sludge formation. The latter forms an ice sheet with a complex layered structure and can significantly complicate the construction and maintenance of engineering structures on the shelf. In most cases, the ice cover of the shelf zone is formed by ice of congelation, infiltration, and intrawater formation, differing in structure and physical characteristics.

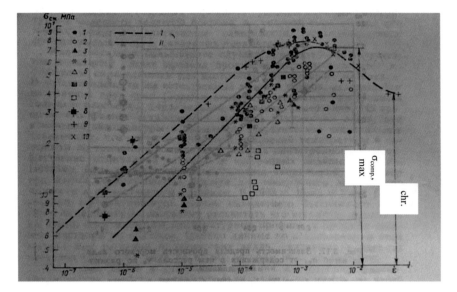

Fig. 3.16 Dependence of the limit for strength of freshwater (*I*) and sea (*II*) ice on the compression from deformation rate, (I) 1—granular ice with a thickness of less than 2 mm; 2—granular ice with a thickness of 2–3 mm; 3—columnar ice thickness 10 mm; 4—columnar ice thickness 11 mm; 5—columnar ice thickness 15 mm; salinity of ice 4–5^0/$_{00}$ at temperature −10 °C; 6—granular ice by other data; 7—columnar ice with salinity 4–5^0/$_{00}$ at temperature −11 °C; (II) 8—columnar ice with salinity 4–5^0/$_{00}$ at temperature −11 °C by other data; 9—columnar ice with salinity 4–5^0/$_{00}$ at temperature −10 $^\circ$C by other data; 10—snow ice

Seasonal changes under the influence of thermometamorphic processes aggravate these differences in the ice cover and lead to significant changes in its strength characteristics. To assess these dependencies, data on seasonal variations in the weighted average temperature and salinity of the ice sheet, and consequently, the volume of its liquid phase, and the functional relationship of these parameters are needed [27].

As it is known, in most cases, when processing test results, the calculation of the ice strength limit is performed according to formulas, valid for an isotropic homogeneous material. In reality, the natural ice cover, located on the border of two media (water and air) is isotropic only in a plane, parallel to the freezing plane. Its physical and mechanical properties change due to changes in temperature and salinity. Theoretical studies show, that, taking into account the profiles of temperature and salinity changes, the strength properties of anisotropic ice differ significantly from the properties of isotropic ice plates [100–102].

Therefore, when destructive stresses are found, for example, during bending, the use of formulas for isotropic ice can lead to significant errors. In 1983, y. Cox and Weeks have created a mathematical model of annual snowless sea ice, which allows us to calculate its true bending strength, if temperature profiles and salt solution profiles are known by thickness. However, data on the use of this model for solving

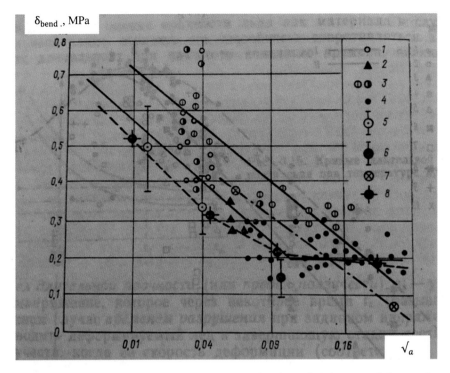

Fig. 3.17 Dependence of the limit for bending strength of sea ice on the brine content in it, according to various data

practical problems, related to the assessment of the bending strength of ice are not yet known.

The above-mentioned with regard to the effect of the anisotropy of the physical and mechanical properties of natural ice on its bending strength gave the reason to IAHR committee, dealing with problems, related to the study of ice strength, to consider the value of $\sigma_{bend.}$, obtained, when testing cantilever beams in full-scale conditions, only as a parameter, necessary to describe the properties of ice, related to bending strength, and not as the true value of the strength limit of the ice field.

Figure 3.18 shows the calculated characteristics of seasonal variability of sea ice bending strength. The calculation was made, taking into account generalized data on the weighted average monthly values of temperature and salinity of six ice types of different ages. The strength values are reduced to the conditionally instantaneous strength of a cantilever beam afloat, cut from ice of average thickness, corresponding to age gradation, without taking into account the direction of application of the load.

As can be seen from Fig. 3.18, the greatest convergence of the results corresponds to the calculated data, obtained on the basis of measurements $\sigma_{bend.} = f(c_l, T)$ and calculation $\sigma_{bend.}$ by the volume of brine in ice. In addition, the obtained results indicate, that under the influence of thermometamorphic processes, the strength of the ice cover up to 1 m thick in summer decreases 5–10 times, compared to its

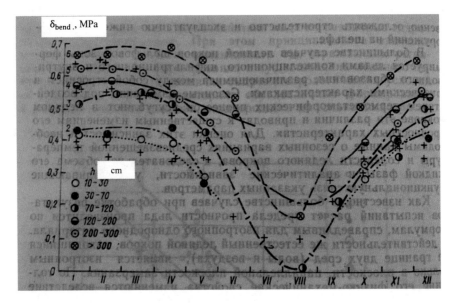

Fig. 3.18 Seasonal variability in the strength of sea ice of various thicknesses. The «+» sign indicates the corresponding assessments $\sigma_{bend.} = f(c_l)$

strength, in winter. This circumstance requires taking into account the spatial and temporal variability of the structure and physical properties of ice in solving a number of scientific and practical problems of ice science and ice technology [103].

Below we will briefly consider the range of variability of the characteristics of the compressive strength of ice. As it was shown above, the destruction of ice is not caused unambiguously by the values of certain stress limits. Due to creep, the beginning of ice destruction and the value of internal stress, corresponding to this moment, significantly depend on the speed of application of the load, deformation conditions and other factors.

This is the reason for the large range of limits, determined by various researchers. For example, numerous experiments have established, that, in addition to increasing the strength with a decrease in temperature, the compressive strength of ice is greater, if the load in it is applied perpendicular to the directions of the axes of the crystals, compared to the strength of the same ice, but in which the force is directed along the axes of the crystals. Compression of ice samples along the crystals leads to the formation of separation cracks between them. When compressed across the axes of crystals, shifts along the basic planes are possible. Experiments show that the limits of ice strength for uniaxial compression at different directions of compressive forces may differ by 3–4 times [103].

According to the generalized experimental data on the influence of the deformation rate, the maximum strength of ice samples corresponds to the deformation rates $\dot{\varepsilon}_p \approx 10^{-3}$ s^{-1}, i.e. with a fracture time, not exceeding several seconds. This strength value corresponds to the moment of transition from semi-brittle to brittle destruction.

The term «transition zone» from plastic destruction to brittle one occurs more often in the literature, as can be seen from Fig. 3.16.

The results of the uniaxial compression test of samples, taken from different ice horizons, give different values of strength limits [104]. The question arises, how to determine the compressive strength of ice for ice «boar», having the limit of strength, obtained from the results of tests of small samples. Studies, conducted in Canada, have confirmed the assumption, that the uniaxial compression strength of a «boar» of ice with a thickness, equal to the thickness of the ice sheet, can be defined as the weighted average strength from tests of small samples, taken of temperature distributions and ice cover under natural conditions. Tests for compression of a «boar» of columnar-granular ice with a thickness of 1.3–1.6 m at a deformation rate $\varepsilon^{\cdot}_{p} \leq 10^{-3}$ s^{-1} and a comparison of the calculation for strength of such a «boar» by testing small samples from the same ice showed satisfactory convergence of the results. When changing $\sigma_{comp.}$ of small samples from 1 to 3.85 MPa, the calculated strength differed by 30% from the experimental one.

Information about the effect of size on the strength of ice is covered very poorly in the literature. There is the following empirical relation for the uniaxial compressive strength of freshwater ice $\sigma_{comp.}$ (MPa) [102]:

$$\sigma_{comp.} = 1.3 \, k_i h_i^{-0.2} \tag{3.28}$$

where $k_i = 1$ at an average ice temperature, equal to -4 °C, h_i—ice thickness, m.

At a standard ice test temperature, equal to -10 °C, and the grain size is 0.4–1.0 mm, $\sigma_{comp.}$ increases by about 30%, compared to $\sigma_{comp.}$ at -4 °C.

Accordingly, the above equality takes the form

$$\sigma_{comp.} = 1,7 \, h_i^{-0,2}. \tag{3.29}$$

Under the conditions of lateral loads, represented by the interaction of ice with piles, the following relation was experimentally obtained [102]:

$$\sigma_{ip} = \sigma_{comp.} \cdot 0,57 \left(\frac{D}{h_i}\right)^{-0,5} h_i^{-0,4} \tag{3.30}$$

where D—pile width, σ_{comp}—the own strength limit of a small sample for uniaxial compression at $T_i \approx 0$ °C and $\varepsilon^{\cdot p} = 10^{-3}$ s^{-1}. The «scale effect» does not manifest itself in uniaxial compression tests, $D/d_{er} > 25$, where d_{er}—the size of the crystal. With uniaxial compression at $T_i = -10$ °C, $\sigma_{comp.}$ approximately in 1.6 times higher than the corresponding value at 0 °C. In the case $T_i = 0$ °C and $d_{cr} = 0.4$ mm this value can be put equal to 3.5 MPa, when $D/h_i = 1$ and at the same time we get the relation

$$\sigma_{ip} = 3,2 \, h_i^{-0,4}. \tag{3.31}$$

Having performed the comparative analysis, described above, and putting $h_i = 0.15$ m, we obtain the following values of limit for strength of the uniaxial compression and indentation resistance [102]:

$$\sigma_{comp.} = 2,5 \text{ MPa и } \sigma_{ip} = 7,0 \text{ MPa}$$

Taking into account, that the relation $\sigma_{ip}/\sigma_{comp.} \approx 3$, it can be assumed that these values are quite consistent.

Regardless of the main reasons for the decrease in strength with an increase in size, the scale effect in ice can be expressed as a function σ_{ip} from the value of $a^{-0,6}$ for areas of contact up to approximately $a = 1.25$ m [103–105].

3.6 Seismoacoustic Properties of Ice, Used to Solve Problems of Civil Aviation

It is known that elastic waves, propagating in a medium, carry information about their physical properties. For example, when studying the mechanics of ice sheets, such parameters of wave processes as their propagation velocities, absorption, and scattering characteristics become important [105]. They can be used to easily calculate the elastic modulus, Poisson's coefficient, viscosity, and in some cases, to estimate the strength of ice during compression and bending.

3.6.1 Types of Elastic Waves in Ice Sheets and Characteristics of Their Propagation

As in any solid body, a wide variety of types of elastic waves can be registered in ice (depending on its shape and size): from volumetric longitudinal and shear waves, characteristic of a medium, whose dimensions are many times greater than the wavelength of the recorded elastic process, to surface (Rayleigh, bending-gravity, air-connected, etc.), when the wavelength is much larger than the characteristic size of the ice formation under study (whether it is a sample of limited dimensions, an idealized infinite plane-parallel plate of the sea ice cover, a glacier, etc.). There is reason to believe that the field of knowledge about waves in ice as a material and in snow-ice natural formations is currently quite saturated with experimental data and may well become the subject of a separate monographic review. Therefore, here we will only briefly focus on some information that, in our opinion, will give an idea of the range of variability (including seasonal, if we consider the annual cycle of measurements, for example, the propagation velocities of ultrasonic and seismic waves in the sea ice cover) the parameters of natural wave processes, that can be used in the consideration of remote seismoacoustic methods for studying sea ice sheets.

Figure 3.19 presents generalized data on the temperature dependences of the propagation velocities of longitudinal and transverse ultrasonic waves. The areas of value of velocity for single crystals, freshwater, and sea polycrystalline ice are indicated by the corresponding hatching. The maximum and minimum values of these regions correspond to different values of crystal orientation, their sizes, the content of air inclusions and brine, i.e. those basic parameters, from which, in addition to temperature, the velocity depends. Since the velocity measurements were made with an accuracy of at least 1%, attention is drawn to the high sensitivity of the acoustic method to the state of the object under study, i.e. to temperature, salinity and ice structure.

A very indicative picture of the influence of density and structure on the speed of sound in the medium under study can be observed by the example of the vertical distribution by velocity in the air—snow—sea ice cover–sea water system. Figure 3.20 shows a wide velocity maximum in the densest layers of ice and two small narrow minima at the boundaries of the transition from the snow-ice surface to the air and in the zone of the so-called openwork transition layer from ice to water.

Figure 3.21 presents experimental data, characterizing the seasonal variability of wave propagation in the sea ice cover. An idea of the integral values of the propagation

Fig. 3.19 Generalized temperature dependences of the propagation velocities of longitudinal (C_l) and transverse (C_t) ultrasonic waves in ice. 1—single crystals; 2—freshwater polycrystalline ice; 3—sea long-term ice; 4—sea annual ice of medium thickness

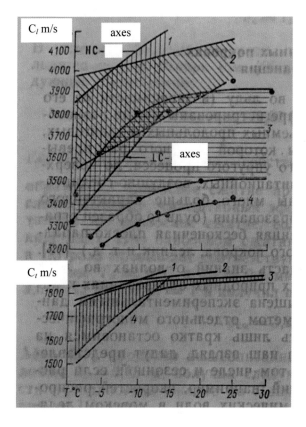

Fig. 3.20 Schematic distribution for vertical of the speed of sound in the air-snow-ice cover-sea water by salinity $34^0/_{00}$. System, C_{air} and C_w—the speed of ultrasonic waves in air and water, respectively

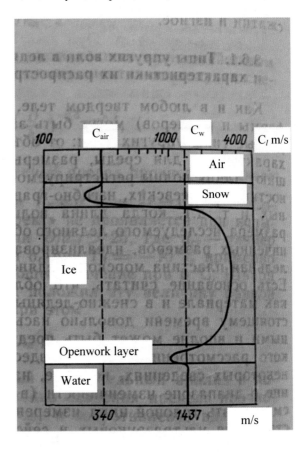

velocity of seismic waves and the velocity of bending-gravity waves in the sea ice cover can be obtained from the data in Table 3.15 and Fig. 3.22.

In the early 70s, fluctuations of ice cover particles with a period from units to several tens of minutes (wavelength approximately 300–1500 m, amplitude 1–5 mm) were detected in the Arctic Ocean. These waves propagate at a phase velocity of 0.5–2.0 m/s. Experiments and theoretical studies have confirmed that the characteristics of such unusually slow waves are in good agreement with existing ideas about the characteristics of internal short-period waves [106]. It is assumed that internal waves, which are own vibrations of layers of a stably stratified sea, reach the surface, causing fluctuations in the ice cover with the same period, phase velocity, and direction of propagation. The lifetime of *slow waves reaches several hours*.

The propagation of acoustic waves in ice is accompanied by attenuation, i.e. a decrease in their amplitude. Attenuation is caused by the absorption of sound by the environment (irreversible transfer of sound energy to other types), divergence of the wave front, scattering on obstacles, re-emission of waves into the environment, and other reasons. It is known that the amplitude of pressure P in a harmonic wave decreases exponentially

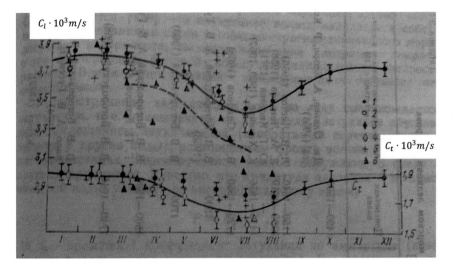

Fig. 3.21 Seasonal changes in the weighted average (by the thickness of the layer of long-term sea ice 0.5–3.0 m) values of wave velocities (C_l and C_t) as a result of statistical processing, obtained by pulsed ultrasonic method at drifting stations in comparison with the results of seismic studies. (1) CP-13; (2) CP-20; (3) CP-4; (4) CP-10; (5) data of V.N. Smirnov and Yu.G. Kiselev; (6) data of K. Hankins

$$P = P_0 e^{-a_{at} t}, \tag{3.32}$$

where a_{at}—the time coefficient of attenuation by amplitude (the coefficient of attenuation by power is equal $2a_3$). Its dimension coincides with the frequency dimension.

The amplitude of pressure in a plane wave decreases with distance, according to the law [78]

$$P = P_0 e^{-(a_{at}/cs)x} = P_0 e^{-\beta x}, \tag{3.33}$$

where P_0—pressure at the starting point $x = 0$, and c_3—the speed of sound. The value $\beta = a_{at}/c_3$ is called the spatial coefficient of attenuation. Its dimension is the same as that of the wave number: $[\beta] = [m^{-1}]$.

Depending on the expected value of attenuation, methods for determining either a_3 or β are used in practice. At high frequencies, when a plane wave can be created in a highly absorbing medium, β is determined. To do this, the amplitudes of sound pressure P_1 and P_2 are measured at two points at a certain distance Z along the line of sound propagation, then

$$\beta = \frac{1}{L} \ln \frac{P1}{P2} \tag{3.34}$$

or, if $A = P_1 / P_2$, then β can be expressed in decibels:

Table 3.15 Velocity (m/s) of propagation of seismoacoustic waves in the sea ice cover, according to experimental data

Type of ice	Temperature of ice (°C)	Measurement method	Longitudinal wave	Transverse wave
Thick annual	−7, …, −8	Pulse seismoacoustic method on bases, equal to several tens–hundreds of meters	3010–3490	1490–1560
Annual of medium thickness	−1, …, −10		2500–3300	1150–1640
			2600–3540	1230–1810
			2640–3220	
Long-term	−1, …, −14		2997–3881	1419–1856
Long-term ($H = 4.5$ m)			3590–3900	1720–1850
Annual (May) $H = 0.2$–1.9M	−5	Pulsed ultrasonic method, sounding of small samples	3000	1500
Long-term (May) $H = 0.9$–3.0 m	−5		3620	1700
Long-term $H = 0.3$–1.3 m	−1, …, −20		3200–3900	1500–1900
Long-term $H = 0.4$–2.0 m	−1, …, −30		3580–3980	1740–1940
Long-term $H = 0.3$–3.0 m	−5, …, −25		3300–3700	
Annual $H = 0.4$–1.4 m	−5		3650	

$$\beta = \frac{A}{8.68L}. \tag{3.35}$$

For low frequencies and weakly absorbing media, when a very large segment L is required for a sufficiently accurate measurement, the method is not used. In such cases, the time coefficient of attenuation a_3 is measured. In coarse-crystalline fresh ice, attenuation is determined by reflection at the crystal boundaries and weakly depends on frequency. Frequency dependence β in the upper and middle desalinated layers of sea long-term ice has the form $\beta = c_1 f + c_2 f^4$, where $c_1 = 4.8 \times 10^{-2}$ dB/(m · κHz), $c_2 = 0.4 \times 10^{-4}$ dB/(m · κHz). For the lower layer of long-term ice and frozen water at a temperature of about −18 °C and in the range of 0.3–0.7 MHz, the type of frequency dependence remains, but the role of the quadratic term increases. The obtained results show that Rayleigh scattering on air inclusions with sizes significantly larger than the wavelength does not make a noticeable contribution to the

Fig. 3.22 Dispersion curves of phase and group velocities of bending-gravitational waves in the ice cover. Thickness of ice: 1—0.1 m; 2—1 m; 3—3 m; 4—10 m; 5—100 m. Sea depth 3000 m

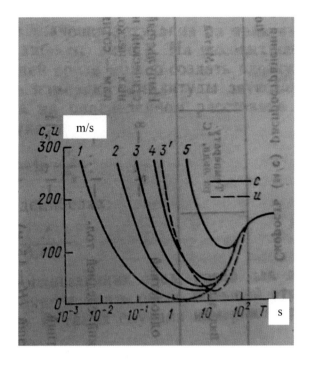

overall mechanism of attenuation of sound waves in ice (Fig. 3.23). A measurement β was carried out in the ice cover in the frequency range of 10–500 кHz. His experimental points fit pretty well on a straight line of the form $\beta = c_1 f + c_2 f^4$, where $c_1 = 4.45 \times 10^{-2}$ dB/(m · кHz), $c_2 = 2.18 \times 10^{-10}$ dB/(m · кHz).

Fig. 3.23 Attenuation of seismoacoustic waves β in ice. 1—large-crystal lake ice at $T = -5$ °C; 2-annual sea ice at $T = -6$ °C; 3—desalinated layers of long-term ice at $T = -19$ °C; 4—Gy-white ice (frozen water) at $T = -19$ °C; 5—the bottom layer of long-term ice at $T = -17$ °C; 6—the average layer of annual ice of average thickness; 7—freshwater ice

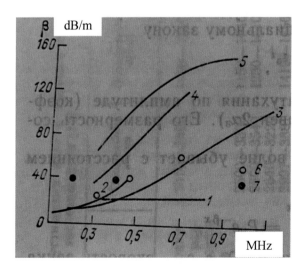

Sound attenuation in young sea ice is mainly caused by viscous losses due to the presence of brine in the ice pores. The coefficient of attenuation β, according to the empirical dependence, obtained by them, has the form $\beta = 7 \sqrt{f}$ (here β—in dB/m, f—in кHz), it is valid for the range 60 Hz $< f < 5 \times 10^5$ Hz. The validity of the ice model as a porous structure with a skeleton of low-compressible ice and pores, filled with brine, is confirmed by a satisfactory coincidence β of the experimentally obtained data with the calculated ones. A significant attenuation of elastic waves in the ice structure can be caused by the peculiarities of ultrasound scattering by its lower surface. When an ultrasonic wave falls on the lower surface, a very intense scattering of ultrasound is observed, which can be recorded in an aqueous medium at various angles, including angles far from the direction of the mirror reflection [81]. From experimental scattering indicators for frequencies of 100 and 200 kHz, it follows, that plane–parallel ices, but with a thickness of 0.02–2.0 m, represent the properties of a diffuse diffuser in both normal and inclined incidence. In addition to diffuse scattering, a mirror reflection is also clearly observed, which returns at least 20% of the power of the incident ultrasonic wave to the aqueous medium.

3.6.2 Spectra of Natural Oscillations

The possibility of using the seismoacoustic method to study the stressed state of the ice cover from the occurrence of low ice pressure to the stage of its destruction is due to the effect of increasing stresses in the ice. The increasing stresses lead to a violation of the «continuity» of the ice from microcracks to large dynamic cracks, extending several hundred meters. At the same time, energy is released, which propagates in boundary media in the form of elastic oscillations of a very wide range. The parameters of these oscillations (frequencies, displacement amplitudes and sound pressures, polarization and wavelengths, variability of their propagation velocities, pulse durations and their density over time) can be objective criteria for determining the general state of the ice before and at the time of the violation, and in some cases allow us to judge the details of the studied phenomena [99–101].

When studying the physics of sound formation, accompanying the process of ice destruction, both in laboratory and in full-scale conditions, interesting data were obtained on the spectral and scale-energy characteristics of sound and seismoacoustic signals, arising at various stages of the stressed state of ice under the influence of dynamic and static loads.

The destruction of ice in reservoirs during its interaction with structures, during hummocking, etc. is a complex dynamic process. Observations show that separate processes can be distinguished from this complex phenomenon, such as the formation of dynamic and thermal cracks, crushing of individual fragments, ice friction on ice, etc. The signals, recorded during the destruction of ice, are pulses of various amplitudes and durations. Pulse noise spectra, corresponding to various mechanisms of sound emission by ice, have their own characteristic features. The presence of pronounced maxima in the spectra may indicate the mechanism of sound emission by ice fragments, oscillating with their inherent natural frequencies. The destruction of ice in a liquid can also be accompanied by cavitation phenomena [106].

One of the important conclusions of this kind of research is the possibility of choosing the optimal operating frequency range of recording acoustic equipment. Thus, the frequency band 100–400 Hz was chosen as the most characteristic for recording various kinds of dynamic processes, occurring in the ice cover and associated with its destruction. When comparing changes in the level of acoustic emission of ice due to the occurrence of temperature stresses in it, exceeding the ice breaking limit, it is necessary to be guided by the fact that cracks, creating noise, can occur only at a well-defined rate of change in air temperature. For example, at the rate of temperature change of 1 °C/h and its continuous decrease for 12 h from −16 to −26 °C, the stress in the ice reaches its limit of strength for breaking after 6–7 h. A slow change in temperature (0.4 °C/h) leads to the achievement of the limit of strength only after 20–24 h, if the temperature drop does not stop during this time. Slower temperature changes do not cause thermal cracking, because during this time the plastic properties of ice will take effect.

In the Arctic basin, the wave spectrum is characterized by a relatively uniform increase in the amplitude of displacement of vertical vibrations of ice particles with periods of 0.1–60 s in deep water and 0.1–100 min on the shelf. The upper limit of the velocity of waves, having a period of 40 s, is 37.3 m/s. The spectra of the recorded waves in the ice sheet with a thickness of 3 m are shown in Fig. 3.24.

It has been experimentally established that the relationship between wind and bending-gravitational fluctuations of ice is significantly manifested only at a wind

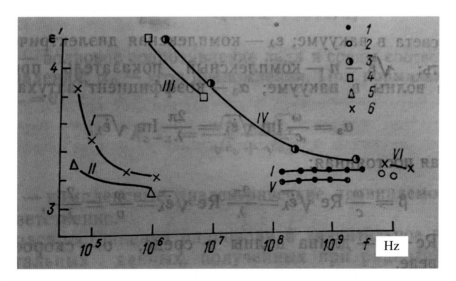

Fig. 3.24 Dependence of ice permittivity on frequency for different temperatures, according to experimental data

speed of 10 m/s or more. In this case, the so-called resonance effect of wave propagation velocities may manifest itself, which consists in the fact, that at a critical
value of wind speed, equal to the minimum speed on the corresponding dispersion
curve, waves are excited more efficiently in the ice cover. The ice belt is an effective
low-pass filter. Thus, waves with a period of 4 s disappear from the spectrum as soon
as measurements begin to be carried out in the ice zone.

The energy of the waves with a period of 8 s decreased, but was still noticeable at
the size of ice floes up to 40 m. There is no noticeable energy loss for waves with a
period of 16 s or more. These waves are given in large (reaching several kilometers
across) ice fields with an ice thickness of up to 2 m. Near the coast of Antarctica in
the Davis Sea, the waves had a period of 16–25 s, despite the fact, that they traveled
a distance of about 1500 km from the place of origin in the zone of «violent» forties
latitudes. With the thickness of the ice cover near the shore of 2.5 m, the amplitude
of waves with a period of 25 s reached 5 cm.

3.7 Electrical Properties of Ice, Used to Solve Problems of Civil Aviation

3.7.1 Speed of Electromagnetic Waves, Characteristics of Their Absorption and Reflection

In ice as an absorbing medium for a plane electromagnetic wave, the relation [102]
takes place

$$E^*(x) = E_0 \cdot e^{-\gamma x} \tag{3.36}$$

where $E^*(x)$—strength of field; E_0—the complex amplitude of the wave at $x = 0$;
γ—the constant of propagation at $\mu = 1$ (complex magnetic permeability):

$$\gamma^* = i\left(\frac{\omega}{c}\right)\sqrt{\dot{\varepsilon}\lambda\mu} = i\left(\frac{2\pi}{\lambda}\right)\sqrt{\dot{\varepsilon}\lambda} = a_{at} + i\beta \tag{3.37}$$

Where $i = \sqrt{-1}$; ω—circular frequency; $c = 1/\sqrt{\mu_B \varepsilon_B} = 2 \cdot 998 \cdot 10^8$ m/s—the
speed of light in a vacuum; $\varepsilon^*{}_\lambda$—complex permittivity; $\sqrt{\dot{\varepsilon}} - n$—complex refractive
index; λ—wavelength in vacuum; a_{at}—coefficient of attenuation;

$$a_{at} = \frac{\omega}{c}\mathrm{Im}\sqrt{\dot{\varepsilon}\lambda} = \frac{2\pi}{\lambda}\mathrm{Im}\sqrt{\dot{\varepsilon}\lambda}.$$

β—phase constant:

$$\beta = \frac{\omega}{c}\mathrm{Re}\sqrt{\dot{\varepsilon}\lambda} = \frac{2\pi}{\lambda}\mathrm{Re}\sqrt{\dot{\varepsilon}\lambda} = \frac{\omega}{\upsilon} = \frac{2\pi}{\lambda_m}$$

where $\lambda_m = \lambda / \mathrm{Re}\sqrt{\dot{\varepsilon}\lambda}$—wavelength in a given medium, υ—speed of wave in this environment.

The speed υ in this environment is determined by the relation

$$\upsilon = \frac{c}{\mathrm{Re}\sqrt{\dot{\varepsilon}\lambda}}.$$

Absorption of electromagnetic energy N_a (dB) in ice thick h is determined by the relation $N_a = 20 \lg e^{a_{at}h} = 8.68a_{at}h = 8.68\frac{\omega}{2c}\sqrt{\dot{\varepsilon}\lambda}\,\mathrm{tg}\,\sigma h$

Specific absorption N (dB/m)—by relation

$$N = \frac{1}{h}20\lg\frac{E_0}{E_h} = 8.68a_{at}.$$

The propagation time of a plane wave over a distance h will be

$$t = \frac{\sqrt{\dot{\varepsilon}\lambda}\left(1 + \frac{1}{8}tg^2\sigma\right)h}{c}. \tag{3.38}$$

Complex refractive index

$$n^* = \sqrt{\dot{\varepsilon}\lambda} = \sqrt{\varepsilon\lambda(1' - \iota tg\sigma)} = \mathrm{Re}\sqrt{\dot{\varepsilon}\lambda} - i\mathrm{Im}\sqrt{\dot{\varepsilon}\lambda} \tag{3.39}$$

where

$$\mathrm{Re}\sqrt{\dot{\varepsilon}\lambda} = \frac{c\beta}{\omega} = \frac{c}{\upsilon}; \; \mathrm{Im}\sqrt{\dot{\varepsilon}\lambda} = \frac{ca_{at}}{\omega}.$$

The reflective properties of ice are characterized by a coefficient for reflection R_{m-i}, which for the case of the incidence of a flat horizontal polarized electromagnetic wave E_{inc} on the boundary of the medium-ice section, is written as

$$R_{m-i} = \frac{\dot{E}_{ref}}{\dot{E}_{inc}} = \frac{\dot{k}_i \cos a_m - \dot{k}_m \cos a_i}{\dot{k}_i \cos a_m + \dot{k}_m \cos a_i} \tag{3.40}$$

where $\dot{k}_i$ and $\dot{k}_m$—the wave resistance of ice and mzzedium, respectively; a_m—angle of incidence; a_i—angle of refraction. With a normal fall ($a_i = a_m = 0$)

$$R_{m-i} = \frac{\sqrt{\dot{\varepsilon}_m} - \sqrt{\dot{\varepsilon}_\iota}}{\sqrt{\dot{\varepsilon}_m} + \sqrt{\dot{\varepsilon}_\iota}}, \tag{3.41}$$

where $\dot{\varepsilon}_m$ and $\dot{\varepsilon}_i$—complex dielectric permittivity of the medium and ice, respectively.

The theory of radio wave propagation and a significant amount of experimental data on the sensing of glaciers in Antarctica and the Arctic allow us to quantify the total attenuation of the amplitude of signal during the passage of the ice layer in the form of the following relation [103]:

$$N_\Sigma = N_g + N_{refl.} + N_f. + N_{scat.} + N_{ab.} + N_{pol.}, \qquad (3.42)$$

where N_g—geometric losses; $N_{refl.}$—losses due to reflection from the boundary of division; N_f—signal change due to focusing; $N_{scat.}$—losses due to scattering; $N_{ab.}$—losses due to absorption; $N_{pol.}$—losses due to the mismatch of the polarization of the signal, received by the antenna, with the polarization of the receiving antenna.

An emergency in the electromagnetic range can be judged, for example, by comparing the following experimental and calculated data. For example, the absorption of electromagnetic waves at frequencies of 0.1–1000.0 MHz with a double passage of their thickness of fresh ice at temperatures from −60 to −1 °C, 1.00 dB/m, and their propagation speeds, depending on the density of the snow-firn layer and the thickness of the glacier, range from about 160–230 m/mcs. In sea ice in the temperature range −5, …, −40 °C, the speed of electromagnetic waves varies from about 60–170 m/mcs, and the specific attenuation at a frequency of 100 MHz at the same temperatures varies almost 30 times, taking values from units to tens of decibels per 1 m, and with an increase in frequency by an order of magnitude, the specific attenuation it increases greatly, reaching values of several hundred decibels per 1 m, according to experimental data.

The dependences of the calculated modules of coefficient for reflection on the frequency at the air–snow and air–freshwater ice boundaries at $E_{inc} = 1$, obtained, taking into account published experimental data, show, that at 103 Hz < f < 106 Hz, the coefficient for reflection from the air–ice boundary decreases from 0.8 to 0.28, and at f > 106 Hz, the coefficients for reflection from the air–snow and air–ice boundaries practically do not depend on either frequency or temperature.

3.7.2 Electrical Properties of Freshwater Ice Cover

Since in the study of natural processes of freezing sea areas, objects of remote research, in addition to the sea ice cover itself, can also be freshwater congelation ice formations (for example, river ones), as well as glacial origin (ice islands, icebergs), this subsection is separated into an independent one.

Figure 3.25 shows the vertical distributions of electrical parameters, which can be used to quickly estimate the average values of ε'_λ, or tg σ of the ice sheet as a whole for a given idealized temperature distribution.

In practice, as a rule, the temperature change is nonlinear. In this case, the average permittivity ε'_λ must be divided by the average velocity c_i of wave propagation in the ice sheet from the ratio $\varepsilon'_\lambda = (\frac{c}{c_i})^2$, according to the known values ε'_λ and m

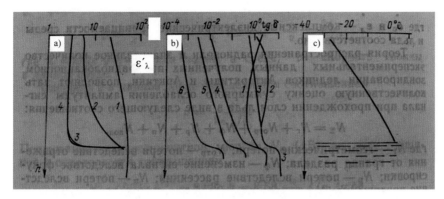

Fig. 3.25 Change in the permittivity and the tangent of the dielectric loss angle in the ice sheet thickness with a linear temperature distribution

a

Curve	1	2	3	4
f Hz	10^3	10^4	10^5	10^6-10^8

Curve	1	2	3	4	5	6
f Hz	10^3	10^4	10^5	10^6	10^7	10^8

homogeneous layers (with low losses) with a thickness of h_k [105]:

$$c_i = \frac{ch}{\sum_{k=1}^{m} \sqrt{\varepsilon'_k k_k}}. \tag{3.43}$$

This ratio also takes into account the change in the density of ice and its textural-structural addition, if known values ε'_λ are substituted into it. Figure 3.26 shows the effect of differences in the texture of ice on its electrical characteristics. With measurement errors ε'_λ and tg σ (1 and 1.5%, respectively) the differences in their values reach up to 30% for ε'_λ and 60% for tg σ.

Earlier, it was noted, that the main factors, determining the electrical properties of freshwater ice cover are its temperature, density and textural and structural composition. If there is dry snow on the ice surface, it will reduce the effective value of ε'_λ and tg σ of such a two-layer system, compared to a homogeneous ice cover. Especially significantly, the electrical properties of the ice cover are changed by snow, containing water. Therefore, the impact of snow should be assessed separately in each case.

The assessment of influence of temperature on the electrical properties of the ice sheet can be made only with a known temperature distribution in its thickness, taking into account, for example, such experimental data as in Fig. 3.5. The electrical parameters in an idealized ice sheet with thickness h with a linear temperature change with a $b = T_0/h$ vary, depending on

$$T_i = T_0 - bh_x \tag{3.44}$$

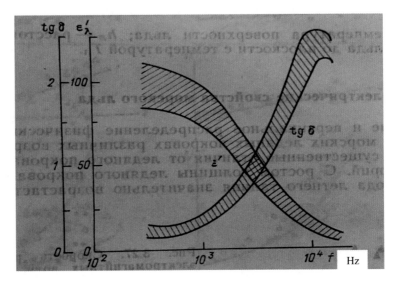

Fig. 3.26 Areas of dependence of the permittivity and the tangent of the dielectric loss angle on the frequency at −10 °C.

where T_0—ice surface temperature; h_x—the distance from the ice surface to the plane with temperature T_i.

3.7.3 Electrical Properties of Sea Ice

The structure and vertical distribution of physical parameters of ice in sea ice sheets of various age stages have significant differences from the ice cover of freshwater water areas. With the increase in the thickness of the ice cover and especially after the period of summer melting, the heterogeneity of the upper layers increases significantly. This is due to the processes of brine runoff, changes in the crystal structure of ice, the formation of a large number of air inclusions, different densities and a number of other factors [87–89].

It has been experimentally established that such heterogeneity causes the presence of a thin electrical structure in sea ice, especially in its upper layers. Due to the presence of brine inclusions, sea ice has a significantly greater attenuation of electromagnetic waves, compared to fresh ice. The great importance of electrical characteristics for the creation and optimization of remote methods for measuring the thickness of the sea ice cover has led to the fact that many authors have studied the electrical properties of these ices in local volumes in laboratories, using various techniques for measurements, and carried out full-scale measurements, allowing to obtain integral characteristics, averaged in thickness. Tables 3.16 and 3.17 and Figs. 3.27, 3.28, 3.29, 3.30, 3.31 and 3.32 present the main results of these studies.

Table 3.16 Refractive index and specific attenuation of Arctic Sea ice in the ultrahigh frequency range (UFR) at changes in temperature and salinity

Temperature (°C)	Wavelength (cm)	Salinity ($^0/_{00}$)					
		2	5	8	2	5	8
Refractive index					Specific attenuation (dB/cm)		
−37	0.78	1.79	1.79	1.79	0.1	0.25	0.42
	3.00	1.79	1.79	1.79	0.01	0.03	0.05
	7.90	1.79	1.80	1.80	0.001	0.002	0.04
−23.5	0.78	1.79	1.80	1.80	0.4	0.9	1.5
	3.00	1.79	1.80	1.82	0.1	0.25	0.4
	7.90	1.80	1.82	1.83	0.002	0.004	0.006
−13.5	0.78	1.78	1.79	1.80	0.7	2.2	4.4
	3.00	1.81	1.85	1.88	0.2	0.55	1.0
	7.90	1.82	1.89	1.95	0.04	0.15	0.3
−5	0.78	1.79	1.84	1.88	1.0	3.0	9.0
	3.00	1.81	1.90	2.10	0.25	1.0	5.0
	7.90	1.82	1.92	3.20	–	0.6	–

3.8 Optical Properties of Ice, Used to Solve Problems of Civil Aviation

3.8.1 Refraction, Absorption, Reflection, and Scattering of Light by Ice

Ice is a uniaxial, optically positive crystal with the property of double refraction; its optical axis coincides with the crystallographic C-axis. It has the lowest *refractive index* of all known minerals (≈0.0014).

The refractive index of ice is a complex value and can be written as [91]

$$\dot{n} = n - i\mu \tag{3.45}$$

where n—the real part of the refractive index, and μ—*extinction coefficient*.

For wave numbers, at which freshwater ice approaches a perfect dielectric, $\mu \approx 0$ and $\dot{n} \approx n$—relative permittivity. At a frequency less than 10^3 Hz $\varepsilon'_\lambda \approx 100$. However, in the Debye dispersion region, the value of ε'_λ rapidly decreases from 3.17 ($n = 1.78$) to its optical value of 1.72 ($n = 1.31$).

The decrease in the intensity of the beam, that has passed through the ice, is a consequence of the absorption and scattering of energy. If I_0 is the intensity of the incident beam, and R is the coefficient of reflection, then at a distance of x from the surface, the intensity will be

Table 3.17 Electrical characteristics of sea ice in the range 100–10.000 MHz

f (MHz)	S_i ($^0/_{00}$)	T (°C)	C_i (m/mcs)	N (dB/m)
100	5.1	−33.0	133	5.2
	5.1	−12.5	93	24.0
	8.2	−22.0	114	20.0
	12.5	−35.0	167	7.0
	12.5	−15.0	101	20.0
	15.8	−35.0	161	6.0
	15.8	−15.0	102	18.0
	35.0	−10.0		51.0
		−30.0		8.0
		−70.0		0.1
	7.0	−10.0		23.0
		−20.0		14.0
		−30.0		2.0
		−40.0		0.8
	3.5	−10.0		10.0
		−20.0		8.0
		−30.0		1.3
		−40.0		0.8
	0.6	−40.0	170	1.0
	0.6	−5.0	170	2.5
	4.4	−40.0	165	1.5
	4.4	−5.0	130	6.0
	8.6	−40.0	165	3.0
	8.6	−10.0	100	8.0
	17.0	−40.0	165	5.0
	17.0	−10.0	40	27.0
	8.0	−10.0		30.0
	3.0	−37.0	171	2.6
	4.5	−32…−35	142	3.9
150	Annual ice (December-January)		60–120	
	March–May		100–120	
	Long-term ice (December-January)		90–130	
150	2.0	−1.0		1.9
		−10.0		0.5
		−20.0		0.3
		−30.0		0.2
300	3.0	−37.0	173	5.2

(continued)

Table 3.17 (continued)

f (MHz)	S_i (%)	T (°C)	C_i (m/mcs)	N (dB/m)
	4.5	−32, …, −35	165	7.1

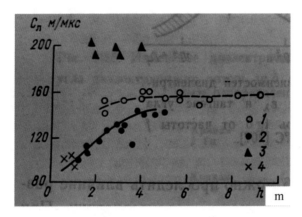

Fig. 3.27 The speed of propagation of electromagnetic waves in sea ice, depending on its thickness. 1—long-term ice without snow; 2—long-term ice with a snow cover of at least 0.5 m thick; 3—ice with a porous texture; 4—annual ice 1–1.5 m thick at air temperature −23, …, −25 °C. The measurements were carried out at a frequency of 200 MHz with pulses of duration of 50 ns

$$I_x = (1 - R)I_0 e^{-ax} \tag{3.46}$$

where a—*ice absorption coefficient*, taking into account the actual absorption and scattering. In most cases, R is a small quantity, so that the pre-exponential term can often be assumed to be equal to I_0.

The absorption coefficient is related to the extinction coefficient by the relation

$$a = 4\pi \mu \bar{v} \tag{3.47}$$

where $\bar{v}$—wave number of radiation, and μ can be calculated by the formula

$$T_{tr} = \exp(-\mu b) \tag{3.48}$$

where the light transmission coefficient through the ice T_{tr} is the ratio of the intensity of the light, transmitted through the ice, to the light, incident on the ice, thickness b.

The extinction coefficient characterizes the total losses during the passage of radiation through the ice sample without taking into account, what happens to this radiation: μ is also called the attenuation indicator.

The absorption capacity is the ratio of the absorbed radiated energy to its full initial value. Consequently, the absorption capacity of an ice plate with thickness b can be expressed by the relation [104]

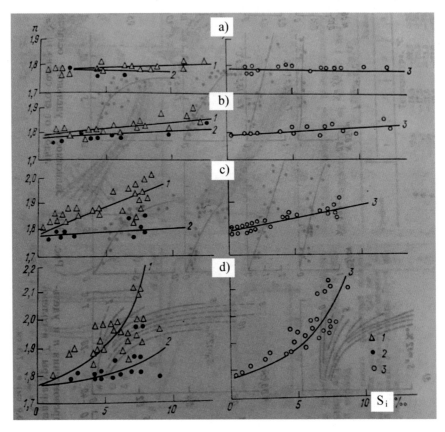

Fig. 3.28 Dependence of the refractive index of radiation by ice on its salinity. Temperature of ice: **a** −37 ± 4 °C; **b** −23 ± 2.5 °C; **c** −13.5 ± 1.5 °C; **d** −5 ± 1 °C; wavelength: 1—7.9 cm; 2—0.78 cm; 3—3 cm

$$a(b) = 1 - e^{ab}. \tag{3.49}$$

For transparent materials, *the reflection coefficient* of a wave polarized perpendicular to the plane of incidence is given in the following form:

$$R_\perp = \frac{\sin^2(\gamma - \gamma')}{\sin^2(\gamma + \gamma')} \tag{3.50}$$

and for a wave, polarized parallel to the plane of incidence:

$$R_\parallel = \frac{tg^2(\gamma - \gamma')}{tg^2(\gamma + \gamma')}, \tag{3.51}$$

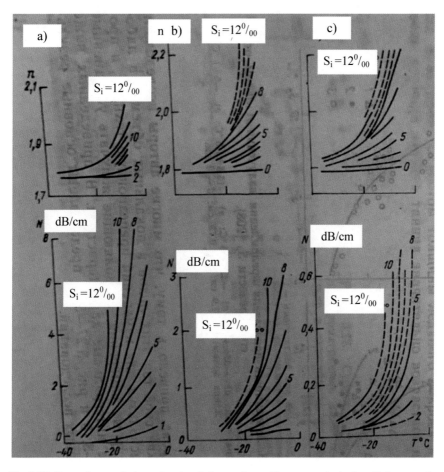

Fig. 3.29 Dependence of the refractive index and specific attenuation of radiation on the temperature and salinity of ice, Wavelength: **a** 0.78 cm; **b** 3.0 cm; **c** 7,9 cm

where γ and γ'—angles of incidence and reflection, respectively. If the wave is not polarized, then

$$R = \frac{1}{2}\left(R_\perp + R_\parallel\right). \tag{3.52}$$

With a normal fall, we have

$$R_{\gamma=0} = \frac{(n-1)^2 + \mu}{(n+1)^2 + \mu}. \tag{3.53}$$

In general case, the absorption coefficient and the reflection coefficient depend on the wave number (wavelength). *The total reflectivity of the surface* (or albedo) of

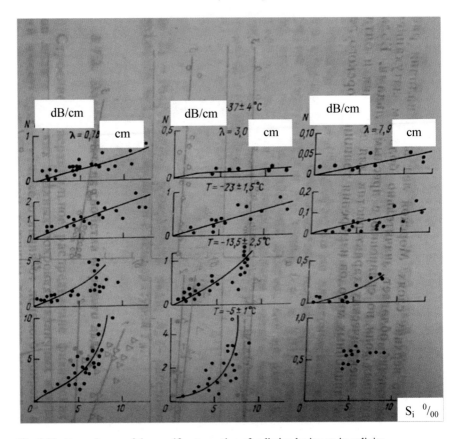

Fig. 3.30 Dependences of the specific attenuation of radiation by ice on its salinity

ice can be expressed in terms of the ratio of the total energy, reflected by the surface, to the total incident energy.

The data in Table 3.18 give an idea of the albedo values of the snow-ice surfaces of the Arctic ice sheet. The scattering properties of ice thicknesses are described by the angular distribution function of the scattered radiant energy, which can be defined as the distribution of the spectral (at some wavelength λ) *scattering index* $\sigma^0(\gamma, \lambda)$ in a given direction (γ is the angle between the direction of the incident beam of light and perpendicular to the face of the elementary scattering volume). It is convenient to write the relative distribution of the intensity of light, scattered by an elementary volume, at the corners, by the function:

$$x(\gamma, \lambda) = 4\pi \frac{\sigma^0(\gamma, \lambda)}{\sigma^0(\lambda)}, \tag{3.54}$$

which is called the indicatrix of scattering.

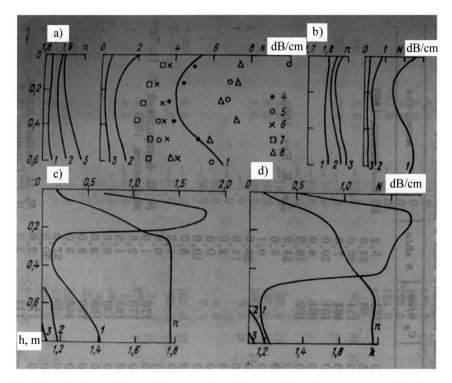

Fig. 3.31 Changes in the refractive index and specific attenuation of radiation by the thickness of the ice cover in the winter-spring period: **a** thin annual winter ice; **b** thin annual autumn ice; **c** two-year ice; **d** long-term ice; wavelength: 1—0.78 cm; 2—3.0 cm; 3—7.9 cm; structure of electrical parameters in winter ice cover: 4–16 January; 5–15 February; 2–6 March; 4–7 April; 8–16 May

The shape of the indicatrix of scattering depends on interference and diffraction phenomena, arising from the interaction of radiation waves with scattering ice particles.

Figures 3.33, 3.34, 3.35 and 3.36 present experimental data.

The analysis of the indicatrix of light scattering by sea ice of various ages has shown that the attenuation of incident radiation depends on the size, shape and concentration of air inclusions, the volume of brine, the orientation of crystalline grains. A comparison of textural and structural photographs and the corresponding ice scattering indicatrix shows that even a barely noticeable difference in ice structure and texture significantly affects the quantitative characteristics of the indicatrix of scattering. The coefficient of transmission of ice significantly depends on the change in the orientation of the crystals with respect to the incident light beam (see Fig. 3.35).

The coefficient has the greatest value, when the beam direction corresponds to the direction of the preferred orientation of the crystals. Then it decreases sharply to a certain level, taking the lowest critical value at an angle of 50° (the angle of refraction

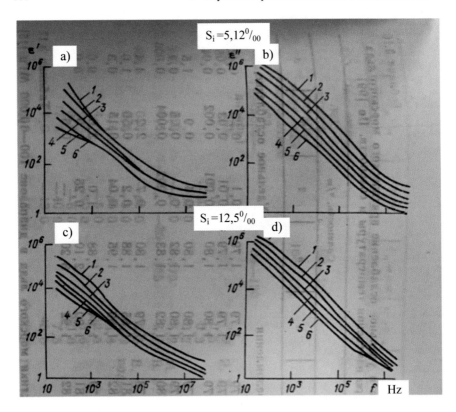

Fig. 3.32 Dependence of the permittivity of sea ice on the frequency at different temperatures

Graph	Curve					
	1	2	3	4	5	6
a	-12,5	-14,5	-22,5	-25,0	-29,0	-33,0
b	-12,5	-17,5	-22,5	-25,0	-29,0	-33,0
c	-15,0	-21,0	-23,0	-35,0	-31,0	-35,0
d	-15,0	-19,0	-23,0	-27,0	-31,0	-35,0

at the ice-air boundary of division). The radiation attenuation index increases sharply with increasing salinity of sea ice (see Fig. 3.34).

3.8.2 Emissivity of Ice

Any physical body at T > 0 K emits electromagnetic waves. Belonging to one or another area of the electromagnetic wave spectrum is produced by the micro-processes, responsible for radiation (internal and external electronic transitions, oscillatory and rotational movements of molecules, lattice oscillation, etc.). Infrared

Table 3.18 Average albedo of snow and ice cover

Character of surface	Destruction of the surface (%)	Albedo
Freshly fallen snow on the ice	0	0.88
Dense snow on ice	0	0.77
Snow on ice at the beginning of melting	0	0.67
Intensely melting snow, sometimes white ice	10	0.62
Intensely melting snow and melting white ice	20	0.56
Melting ice with a white surface	30	0.51
Melting ice with poorly developed snowflakes	40	0.47
Intensely melting ice with snowflakes	50	0.41
Melting ice, completely covered with snowfields	70	0.33

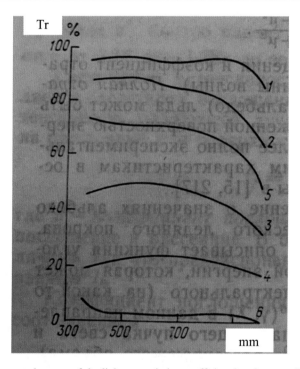

Fig. 3.33 The spectral course of the light transmission coefficient by plane-parallel plates 2 cm thick from some varieties of freshwater and sea ice and snow, 1—pure coarse-crystalline ice with a density of q = 917 кg/m³, the reflection coefficient of which is R = 4%; 2—freshwater ice with tubular air inclusions, q = 895 кg/m³; R = 5–6%; 3—water-snow bubble ice, q = 876 кg/m³; R = 12–16%; 4—surface layer of long-term ice, R = 10–15%; 5—middle layers of long-term ice of columnar-granular structure, q = 910–920 кg/m³, S = 1–6⁰/₀₀, R = 11–14%; 6—snow, q = 280 кg/m³; R = 65–80%

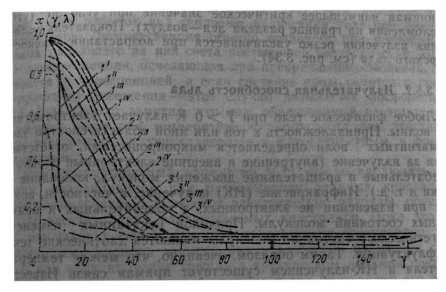

Fig. 3.34 Normalized indicatrix of light scattering by various layers of long-term sea ice, The angle γ between the predominant orientation of the crystals and the direction of the incident beam: 1—0°, 2—20° and 3—50°; I—$l = 3…15$ cm, $q = 895$ кг/m³, S = 0.03⁰/₀₀, R = 19%; II—$l = 15…59$ cm, $q = 912$ кг/m³, S = 0.22⁰/₀₀, R = 17%; III—$l = 59…192$ cm, $q = 917$ кг/m³, S = 0,47⁰/₀₀, R = 15%; IV—$l = 250…282$ cm, $q = 918$ кг/m³, S = 1,36⁰/₀₀, R = 17% (l—thickness of layer, q—ice density, S—salinity of ice, R—coefficient of reflection)

Fig. 3.35 The dependence of the relative light transmission by ice on the orientation of sea ice crystals

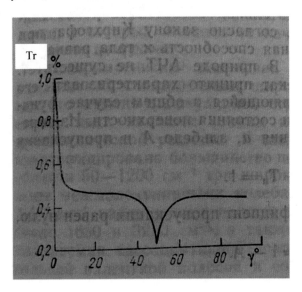

Fig. 3.36 Dependence of the light transmission coefficient of sea ice on its density

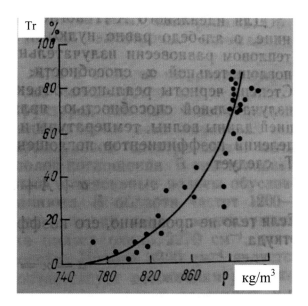

radiation (IR), in particular, occurs, when the oscillatory and rotational states of a molecule, not electronic, change. The reason for this energy state of the molecule is chaotic thermal fluctuations. Thus, it is obvious that there is a direct relationship between body temperature and IR radiation. The well-known Planck equation establishes the relationship between the intensity E_r of the radiation of the so-called a completely black body (CBB) and its temperature T for any range of wavelengths λ [106]:

$$E_r = \int_{\lambda_1}^{\lambda_2} \frac{2\hbar c^2}{\lambda^5} \frac{1}{e^{\frac{\hbar c}{\lambda KT}} - 1} \quad (3.55)$$

where $\hbar$ and K—Planck and Boltzmann constants, c—the speed of light in a vacuum.

For an ideal CBB, the absorption capacity is one, and the albedo is zero, and, according to Kirchhoff's law, at thermal equilibrium, the emissivity of a body is equal to its absorption capacity a_{at}. CBB does not exist in nature. The degree of blackness of a real object is usually characterized by its emissivity, which is generally a function of wavelength, temperature, and state of surface. From the definitions of the absorption coefficients a, albedo A and transmission T_i follows [106]

$$a = \kappa = 1 - A. \quad (3.56)$$

Since the absorption capacity of ice is large in the wave number range below 10,000 cm^{-1}, and the reflectivity is small, the radiation of ice in the infrared range should be close to unity. And, indeed, the experimentally obtained values of κ are 0.97.

Under certain conditions, ice also has *luminescent properties*, i.e. the ability to emit light, which is an excess over the thermal radiation of the body at a given temperature. Luminescence usually occurs, when bodies absorb light of sufficient intensity, that incident on them. Moreover, the luminescence, that disappears, when the lighting stops, is called fluorescence; and, if the glow continues after the lighting stops, this case is called phosphorescence.

Triboluminescent effect in ice can be observed during friction, crushing, or splitting of ice crystals. *Triboluminescence* is caused by electrical discharges, caused by electrification of fractured crystals. So far, this interesting phenomenon has not been systematically investigated by anyone, although by now there are a number of observations of light emission at night, when cracks occur in the Arctic ice cover.

Thermoluminescence of ice occurs, when it is heated after being excited by light or hard radiation, for example, *X*-rays. The peak brightness of thermoluminescence depends linearly on the dose of preliminary irradiation.

Some types of interaction of electrons with ice are accompanied by absorption spectra of trapped ice electrons, very similar to the spectra of hydrated electrons in water and electrons in alkaline ice. The formation of electron capture areas during irradiation is caused by the decay of some water molecules in the ice structure. This decay leads to the formation of vacancies, occupied by electrons. The areas, in which there are trapped electrons, are called *color centers*.

Electromagnetic radiation of ice in the area of infrared spectroscopy is widely used to study its atomic and molecular structure, since the oscillatory spectrum of ice is thin (an indicator of all changes in its crystal lattice). Most of the absorption bands can be identified in the infrared spectrum. In the spectral region of $50–1200$ cm^{-1}, three broad intense bands are caused by intermolecular oscillations. In the frequency range $1200–4000$ cm^{-1}, ice absorption bands with maxima of about $1650–3220$ cm^{-1} are observed, as well as a band of about 2270 cm^{-1}, often called the «associative band». The 3220 cm^{-1} band is a strong valence band and corresponds to the stretching movements of the O-H bond and is close to the frequencies of the O-H valence modes of water vapor. The band with a maximum of about 1650 cm^{-1}, probably, corresponds to the deformation oscillations of H-O-H and is also called the band of bond bending.

In the study of the infrared spectrum area of $50–360$ cm^{-1}, corresponding to inhibited broadcasts, in addition to a large maximum at 229 cm^{-1}, a less intense maximum at 164 cm^{-1} is observed. Peaks at 229 and 164 cm^{-1} can be attributed to the maximum values of the density of oscillatory states due to transverse optical and longitudinal acoustic oscillations, respectively, and part of the spectrum near 190 cm^{-1}—to the maximum of longitudinal optical oscillations. The peak of about 65 cm^{-1} corresponds to the maximum of transverse acoustic oscillations.

3.9 On the Anisotropy of the Physical Properties of Ice

Anisotropy is a phenomenon, consisting in the fact, that the physical properties of a body are different in different directions. Natural ices with different crystal structures, in particular, exhibit anisotropy of mechanical properties to a large extent. For relatively pure polycrystalline ice, two types of «initial» anisotropy are distinguished [77]: (a) *textural anisotropy* due to the shape and location of crystals (or the tortuosity of the boundaries, adjacent to each other) and (b) *structural anisotropy*, associated with the presence of a preferred orientation of the crystallographic C-axes. In contrast to the «initial» anisotropy, that exists in ice at the initial moment of time, in the process of deformation due to recrystallization, manifested in shifts and rotations of individual crystals, *induced anisotropy* can develop in ice.

Textural anisotropy manifests itself well at high temperature, when easier slippage of crystals, relative to each other, is possible. Structural anisotropy is significant in the case, when shear stresses are directed along the predominant direction of the reference planes, if this is the case in ice.

Researchers have established the presence of a difference in the values of the elasticity and strength characteristics of natural ice, depending on the direction of the force, relative to the main optical axis (in single crystals) or the location of structural elements (in columnar granular or fibrous structures) long ago. For example, studies at the drifting station «North Pole-2» have established, that with all types of deformation of sea ice, its strength under load, applied perpendicular to the freezing surface is greater than under load, applied in parallel. This difference is 40–50% for compression, 2–7% for bending, 22–25% for cutting, and 0–25% for impact loading.

The transversal anisotropy of the sea ice cover is quite strongly manifested, when determining the Poisson's coefficient by static methods, especially at a low rate of application of the load to the ice. In these cases, it seems more correct to talk not about Poisson's coefficient, which is true for elastic deformation areas, but about the deformation coefficient or the ratio of the corresponding deformations.

According to theoretical and experimental data, for freshwater columnar-granular and sea annual ice of a fibrous structure with a salinity of 5‰ at a temperature of $-20\ ^{\circ}$C, the transverse deformation v_{yx}, perpendicular to the length of the fibers («columns») is approximately three times greater than the transverse deformation v_{zx}, parallel to the fibers («columns») at a rate of application of load at 10^{-3} MPa/s and rate of deformation $10^{-7}\ \text{s}^{-1}$.

The deformation coefficient v_{yx} in the plane of loading xy (the horizontal plane of ice cover growth; the z-axis is perpendicular to this plane) decreases monotonically from a value of 0.7 at $\dot{\varepsilon} = 10^{-7}\ \text{s}^{-1}$ to $v_{yx} \approx 0.3$ at $\dot{\varepsilon} = 10^{-2}\ \text{s}^{-1}$, i.e. to a value, that we usually call Poisson's coefficient. However, the deformation coefficient v_{zx}, corresponding to the plane xz, parallel to the length of the elongated elements of the ice structure, increases from about 0.2 to 0.3 in the same load range.

Sea ice of fibrous structures (ice of the B3, B4, B5 type) is widespread both in the Arctic, where there is no obvious influence of freshwater runoff, and in Antarctica. It is noted, that under the action of subglacial currents, aggregate crystals of fibrous structures tend to occupy a position, in which their main optical axis is parallel to the flow. Crystals with an excellent direction of the main optical axis are wedged out, giving way to crystals with a more favorable orientation. As a result, the optical axis is parallel to the flow.

Thus, the entire ice field becomes anisotropic. Moreover, the larger the ice layer is composed of identically oriented crystals, the more pronounced the anisotropy of the ice cover will be. Azimuthally oriented structures are most often found in the middle and lower parts of the ice, when the influence of dynamic factors weakens and unfavorably located crystals wedge out [64].

According to experimental data, when the azimuth angle changes, the specific resistance of the median layers of annual sea ice to the horizontally applied load varies from 4 to 6 MPa. The greater strength of the ice is fixed at the direction of the destructive force along the predominant orientation of the C-axes (the azimuth angle is 0°), the lower strength corresponds to 45°, and at an angle of 90° an intermediate value of ice strength is obtained. Young's modulus at the same sounding angles of ice samples varied within 7–8 GPa.

Annual drifting sea ice has anisotropic properties, also in the UFR range of electromagnetic waves. At the same time, while studying the structure and mechanical properties of sea ice, the anisotropy of their electrical characteristics—specific attenuation of electromagnetic waves and dielectric permittivity was investigated.

During the measurements, ice samples of oriented structures rotated relative to the direction of the intensity vector of electric field at an angle of 180 or 360° with a step of 15 or 30°. Thus, the polarization characteristics of the reduction of specific attenuation and refractive index at four wavelengths of the centimeter range were obtained.

Figure 3.37 shows the dependences of the specific attenuation of electromagnetic waves at various frequencies on the angle of rotation of the vector of electric field, relative to the north-south direction of an oriented sample of annual ice, obtained from the center of an ice field by size of $100 \bullet 50$ m^2.

As can be seen from the figure, the coincidence of the specific attenuation maxima at different frequencies is good, and the spread of the angle values does not exceed 15°. Analysis of the ice structure showed, that the greatest specific weakening occurs in the direction, perpendicular to the integral direction of the main optical axes of fibrous ice crystals, which coincides with the direction of the subglacial flow.

The position of the specific attenuation maxima in samples of various ice taken at other points in the area is also determined by the orientation of the crystals. However, the large variation of these maxima at angles at different frequencies, which in some cases reaches 45–50°, is explained by a decrease in the degree of ordering of ice crystals, caused by the variability of the water flow [104].

It was noted that weak anisotropy of electrical parameters is observed even in long-term and two-year-old ice in layers, that have already passed the stage of summer melting, where the spatial ordering of primary ice has been preserved. This makes it possible to assess the changes, that have occurred in the structure of ice under the influence of thermometamorphism in the spring-summer period. In particular, the

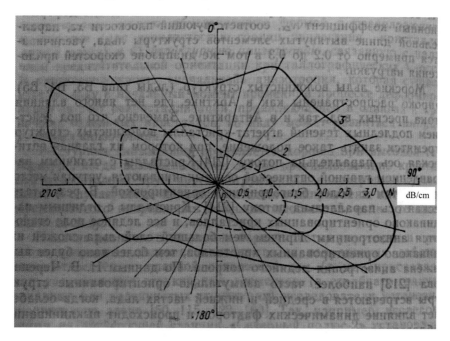

Fig. 3.37 The dependence of the specific attenuation on the angle of rotation of the vector of electric field, relative to the north-south direction of an oriented sample of annual sea ice at different wavelengths. 1—3 cm; 2—1.6 cm; 3—0.8 cm

reversal of the optical axes of ice during recrystallization (the occurrence of induced anisotropy) was noted.

The conducted studies have convincingly proved the presence of anisotropy of the electrical parameters of sea ice and its connection with their crystal structure. In some cases, the difference between the minimum and maximum values of specific attenuation at a wavelength of 0.8 cm reached 5–10 dB/cm. The values of the dielectric constant changed little, no more than 5% from the observed average values.

The conducted studies allowed us to outline the relationship between the crystal structure and the mechanical properties of ice. The latter, as it is known, is the basis for the development of remote sensing methods for ice sheets. Further study of the anisotropy of the electrical parameters of ice can provide prerequisites for the creation of contactless means of determining those directions in the ice, where its strength is minimal.

References

1. Eliseev BP, Kozlov AI, Romancheva NI, Shatrakov YG, Zatuchny DA, Zavalishin OI (2020) Probabilistic-statistical approaches to the prediction of aircraft navigation systems condition. Springer Aerospace Technology, 200 p
2. Akinshin RN, Zatuchny DA, Shevchenko DV (2018) Reducing the effect of the multipath effect, when transmitting information from an aircraft. J Inf Commun 3:6–12
3. Zatuchny DA (2018) Analysis of the impact of various interference on the navigation systems of civil aviation aircraft. J Inf Commun 2:7–11
4. Braykraits SG, Zatuchny DA, Ilyin EM, Polubekhin AI (2017) A methodical approach to determining the place of non-traditional radar methods in promising radar applications. Bull St. Petersburg State Univ Civil Aviat 3(16):72–84
5. Zatuchny DA, Logvin AI (2012) Satellite navigation and air traffic management systems. Study guide. Publishing House MSTU CA
6. Eliseev BP, Kozlov AI, Sarychev VA, Fadeeva VN (2013) Volume V. The theory of electromagnetism from electrostatics to radio polarimetry. Part 8. Electromagnetic waves. Radio electronics. Moscow
7. Shatrakov YuG (2015) Flight safety and the direction of development of simulators for air traffic management specialists. Publishing house of the State university of aerospace instrumentation. St. Petersburg, 516 p
8. Shatrakov YuG (2016) The development of domestic radar aviation technology (encyclopedia). Publishing house Stolichnaya encyclopedia. Moscow, 400 p
9. Automated air traffic management systems. In: Shatrakov YuG (ed). State university of aerospace instrumentation, St. Petersburg
10. Likhgart LP, Kozlov AI, Logvin AI, Avtin IV (2019) Polarization methods for determination and visualization of complex dielectric permittivity in remote sensing issues. Sci Bull Moscow State Tech Univ Civil Aviat 22(4):100–108
11. Kozlov AI, Sergeev VG (1998) Propagation of radio waves along natural routes. Moscow
12. Kozlov AI, Maslov VYu (2004) Differential properties of the scattering matrix. Sci Bull Moscow State Tech Univ Civil Aviat 79:43–46
13. Soloviev YuA (2000) Satellite navigation systems. Moscow
14. Logvin AI, Orlov OE (2002) Satellite navigation and communication systems for air traffic management. MSTU CA, Moscow
15. Bakulev PA, Sosnovsky AA (2011) Radio navigation systems. Radio engineering, Moscow
16. Shatrakov YuG (1982) Angle-measuring radio engineering landing systems. Publishing house Transport, Moscow, 162 p
17. Shatrakov YuG (1990) Macroscopic processing of radio navigation signals. Publishing house of radio and communications, Moscow, 280 p
18. Shatrakov YuG (2018) Development of domestic radar aircraft systems. Publishing house Stolichnaya encyclopedia, Moscow, 400 p
19. Drachev AN, Farafonov VG, Balashov VM (2014) Methods of control of complex-profile surfaces. Radio Electron Issues 1(1):91–99
20. Antsev GV, Bondarenko AV, Golovachev MV, Kochetov AV, Lukashov KG, Mironov OS, Panfilov PS, Parusov VA, Raisky VL, Sarychev VA (2017) Radiophysical support of ultrashort pulse radar systems. In the collection «Problems of remote sensing, propagation and diffraction of radio waves». Lecture notes. Scientific Council of the Russian Academy of sciences on the propagation of radio waves. Murom Institute (branch); Vladimir state university, named after Alexander Grigoryevich and Nikolai Grigoryevich Stoletov, pp 5–21
21. Antsev GV, Bondarenko AV, Golovachev MV, Kochetov AV, Mironov OS, Panfilov PS, Parusov VA, Sarychev VA (2016) Technologies of ultrashort pulse radar of natural environments with high range resolution. Meteorol Bull 8(3):17–22
22. Antsev GV, Bondarenko AV, Golovachev MV, Kochetov AV, Lukashov KG, Mironov OS, Panfilov PS, Parusov VA, Raisky VL, Sarychev VA (2016) Experimental studies of the characteristics of the ultrashort pulse radar. In the collection «Radiophysical methods in remote

sensing of media». Materials of the VII All-Russian scientific conference. Murom Institute (branch) of Federal state budgetary educational institution of higher education «Vladimir state university, named after Alexander Grigoryevich and Nikolai Grigoryevich Stoletov», pp 196–202

23. Ershov GA, Zavyalov VA, Sinitsyn VA (2019) Ways to improve the parameters of radar observability of aircraft by multi-position radar methods. In the collection «Innovative technologies and technical means of special purpose». Proceedings of the eleventh All-Russian scientific and practical conference. In 2 volumes. The series «Library of the magazine "Voenmeh. Bulletin of Baltic state technical university», pp 21–24

24. Balashov VM, Drachev AN, Michurin SV (2019) Methods of control of reflectors of mirror antennas. In the book «Metrological support of innovative technologies». International Forum: Abstracts, pp 44–46

25. Balashov VM, Drachev AN, Smirnov AO (2019) Methods of coordinate measurements in the control of a complex profile surface. In the book «Metrological support of innovative technologies». International Forum: Abstracts, pp 41–43

26. Zatuchny DA (2017) Analysis of wave reflection features during data transmission from an aircraft in urban conditions. J Inf Commun 2:7–9

27. Ivanov YuV, Petukhov SG, Sinitsyn VA (2017) The use of a radar landing system to ensure the landing of aircraft in automatic mode. In the collection «Innovative technologies and technical means of special purpose». Proceedings of the IX All-Russian scientific and practical conference. In 2 volumes. The series «Library of the magazine «Voenmeh. Bulletin of Baltic state technical university». Ministry of education and science of the Russian Federation, Baltic state technical university «Voenmeh», named after D.F. Ustinov, pp 326–330

28. Zatuchny DA, Kozlov AI, Trushin AV (2018) Distinguishing objects of observation, located within the irradiated area of the surface. J Inf Commun 5:12–21

29. Kozlov AI, Logvin AI, Sarychev VA (2007) Radar polarimetry. Polarization structure of radar signals. Radio engineering, 640 p

30. Myasnikov SA, Sinitsyn VA (2019) Features of the construction of a new landing radar. In the collection «Innovative technologies and technical means of special purpose». In: Proceedings of the eleventh All-Russian scientific and practical conference. In 2 volumes. The series «Library of the magazine «Voenmeh. Bulletin of Baltic state technical university», pp 79–84

31. Proshin AA, Goryachev NV, Yurkov NK (2018) Calculation of radio wave attenuation. Certificate of registration of the computer program RUS 2019612561 05.12.2018

32. Proshin AA, Goryachev NV, Yurkov NK (2018) Calculation of the dew point. - Certificate of registration of the computer program RUS 2019612562 05.12.2018.

33. Rassadin AE (2010) The apparatus of atomic functions and R-functions as the basis of mathematical technology for designing radar synthesizing equipment on an air carrier. In: Pupkov KA (ed) Intelligent systems: Proceedings of the Ninth International symposium. RUSAKI, Moscow, pp 224–228

34. Kozlov AI, Amninov EV, Varenitsa YuI, Rumyantsev VL (2016) Polarimetric algorithms for detecting radar objects against the background of active noise interference. In: Proceedings of Tula state university, Technical sciences, no 12–1, pp 179–187

35. Sinitsyn VA, Sinitsyn EA, Strakhov SYu, Matveev SA (2016) Methods of signal formation and processing in primary radar stations. St. Petersburg

36. Yurkov NK, Kuatov BZh, Eskibaev ET (2019) Algorithm of parametric and time control of parameters of aircraft motion management. In: Proceedings of the international symposium Reliability and Quality, vol 1, pp 219–221

37. Abramovich YuI, Spencer NK, Gorokhov AYu (2001) Isolation of independent radiation sources in non-equidistant antenna arrays. Foreign radio electronics. Successes of modern radio electronics, no 12, pp 3–17

38. Baranov AM, Bogatkin OG, Goverdovsky VF et al (1992) Aviation meteorology: Textbook. In: Vasiliev AA (ed). Publishing house Hydrometeo, St. Petersburg, 347 p

39. Aviation radar: Handbook. In: Davydov PS (ed). Transport, Moscow, 223 p

40. Gromov GN, Ivanov YuV, Savelyev TG, Sinitsyn EA (2002) Adaptive spatial Doppler processing of echo signals in the radar air traffic management system. Federal state unitary enterprise «All-Russian scientific research institute of radio equipment», St. Petersburg, 270 p

41. Zhuravlev AK, Khlebnikov VA, Rodimov AP (1991) Adaptive radio engineering systems with antenna arrays. Publishing house of Leningrad state university, Leningrad, 544 p

42. Active-passive radar of thunderstorms and lightning-dangerous foci in clouds. In: Kachurin LG, Divinsky LI. Publishing house Hydrometeorology, St. Petersburg, 216 p

43. Gostyukhin VL, Trusov VN, Klimachev KG, Danich YuS (1993) Active phased antenna arrays. In: Gostyukhin VL (1993). Radio and communications, Moscow, 272 p

44. Drogalin VV, Merkulov VI, Rodzivilov VA et al (1998) Algorithms for estimating the angular coordinates of radiation sources, based on spectral analysis methods. Foreign radio electronics. Successes of modern radio electronics, no 2, pp 3–17

45. Alekseev VG (2000) On nonparametric estimates of spectral density. Radio Eng Electron 45(2):185–190

46. Andreev VG, Koshelev VI, Loginov SN (2002) Algorithms and means of spectral analysis of signals with a large dynamic range. Questions of radio electronics. A series of radar equipment, issue 1–2, pp 77–89

47. UFR antennas and devices (1994) Design of phased antenna arrays. In: Voskresensky DI (ed). Radio and communications, Moscow, 592 p

48. Astapenko PD, Baranov AM, Shvarev IM (1980) Weather and flights of airplanes and helicopters. Leningrad: Publishing house Hydrometeo, 280 p

49. Atlas D (1967) Successes of radar meteorology. Translated from English. Publishing house Hydrometeo, Leningrad, 194 p

50. Bakulev PA, Stepin VM (1986) Methods and devices of selection of moving targets. M, Radio and communications, 288 p

51. Bakulev PA, Stepin VM (1986) Features of signal processing in modern surveillance radar systems: review. Radioelectronics 4:4–20. (News of universities)

52. Banach VA, Werner H, Smaliho IN (2001) Sounding of clear sky turbulence by Doppler radar. Numer Model Opt Atmos Ocean 14(10):932–939

53. Battan LJ (1962) Radar meteorology. Translated from English. Publishing house Hydrometeo, Leningrad, 196 p

54. Bean BR, Dutton ED (1971) Radiometeorology. Publishing house Hydrometeo, Leningrad, 362 p

55. Brylev GB, Gashina SB, Nizdoiminoga GL (1986) Radar characteristics of clouds and precipitation. Publishing house Hydrometeo, Leningrad, 231 p

56. Van Tris G (1977) Theory of detection, evaluation and modulation. Translated from English. Soviet Radio, Moscow, vol 3, 664 p

57. Veiber YaE, Skachkov VA, Smirnov NK (1987) Features of measuring the velocities of relative movements of atmospheric formations by radar methods. Academy of Sciences of the USSR, Moscow, 13 p

58. Vereshchagin AV, Mikhailutsa KT, Chernyshov EE (2002) Features of detection and evaluation of characteristics of turbulent weather formations by onboard Doppler weather radars: Report at the XVIII All-Russian Symposium «Radar research of natural environments» (18–20.04.2000). In the book: Proceedings of the XVI-XIX All-Russian Symposium «Radar research of natural environments», issue 2. St. Petersburg, pp 240–249

59. Vityazev VV (1993) Digital frequency selection of signals. Radio and communications, Moscow, 239 p

60. Vlasenko VA, Lappa YuM, Yaroslavsky LP (1990) Methods of synthesis of fast convolution algorithms and spectral analysis of signals. Nauka, Moscow, 180 p

61. Vostrenkov VM, Ivanov AA, Pinsky MB (1989) Application of adaptive filtering methods in Doppler meteorological radar. Meteorol Hydrol 10:114–119

62. Vostrenkov VM, Melnichuk YuV (1984) Signals of the underlying surface and meteorological objects on the onboard Doppler radar. In: Proceedings of the Central Observatory, issue 154, pp 52–65

63. Gorelik AG (1965) Radar methods for studying atmospheric turbulence. Publishing house Hydrometeo, Moscow, 25 p

64. Gorelik AG, Melnichuk YuV, Chernikov AA (1963) The connection of statistical character- istics of a radar signal with dynamic processes and the microstructure of a meteorological object. In: Proceedings of the Central Observatory, issue 48, pp 3–55

65. Gritsunov AV (2003) The choice of methods for spectral estimation of time functions in the modeling of ultra-high frequency devices. Radio engineering, no 9, pp 25–30

66. Gun S, Rao D, Arun K (1989) Spectral analysis: from conventional methods to methods with high resolution. In the book: Ultra-large integrated circuits and modern signal processing. In: Gun S, Whitehouse H, Kailat T (eds) translated from English, Lexachenko VA (ed). Radio and communications, pp 45–64

67. Jenkins G, Watts D (1971–1972) Spectral analysis and its applications, Translated from English in 2 volumes. Mir, Moscow

68. Doviak R, Zrnich D (1988) Doppler radars and meteorological observations, Translated from English. Publishing house Hydrometeo, Lenngrad, 511 p

69. Dorozhkin NS, Zhukov VYu, Melnikov VM (1993) Doppler channels for radar MRL-5. Meteorol Hydrol 4:108–112

70. Ermolaev VT, Maltsev AA, Rodyushkin KV (2000) Statistical characteristics of criteria in the problem of detecting multidimensional signals in the case of a short sample: Report. In the book: The Third International conference «Digital signal processing and its application»: Reports, vol 1, pp 102–105

71. Zhuravlev AK, Lukoshkin AP, Poddubny SS (1983) Signal processing in adaptive antenna arrays. Publishing House of Leningrad state university, Leningrad, 240 p

72. Zubkovich SG (1968) Statistical characteristics of radio signals, reflected from the Earth's surface. Soviet radio, Moscow, 224 p

73. Ivanov YuV (1990) Asymptotic efficiency of algorithms of space-time processing in coherent pulse radar systems. Editorial board of the Journal Radioelectronics (News of higher educational institutions). Kiev, 12 p

74. Kalinin AV, Monakov AA (2001) Estimation of parameters of weather phenomena, dangerous for aviation by radar method: Report. In the book: The fourth scientific session of graduate students of the State university of aerospace instrumentation, dedicated to the World Aviation and Cosmonautics Day and the 60th anniversary of the State university of aerospace instru- mentation (St. Petersburg, March 26–30, 2001): A collection of reports. Part 1. Technical sciences. St. Petersburg.: Publishing house of the St. Petersburg state university of aerospace instrumentation, pp 236–238

75. Kaplun VA (2004) Radio-transparent antenna fairings. Antennas 8–9(87–88): 109–116

76. Kravchenko NI, Bakumov VN (1999) The limiting error of measuring the regular Doppler shift of the frequency of meteorological signals. News of universities. Radioelectronics 4:3–10

77. Kravchenko NI, Lenchuk DV (2001) The limit accuracy of measuring the Doppler shift of the frequency of a weather signal, when using a bundle of coherent signals. News of universities. Radioelectronics 7:68–80

78. Krasyuk NP, Koblov VL, Krasyuk VN (1988) The influence of the troposphere and the underlying surface on the operation of radar systems. Radio and communications, Moscow, 216 p

79. Krasyuk NP, Rosenberg VI (1970) Ship radar and meteorology. Publishing house Sudostroenie, Leningrad, 325 p

80. Cook Ch, Bernfeld M (1971) Radar signals, Translated from English. Soviet radio, Moscow, 568 p

81. Kulikov EI (1964) Limiting accuracy of measuring the central frequency of a narrow-band normal random process against a background of white noise. Radio Eng Electron 10:1740– 1744

82. Levin BR (1989) Theoretical foundations of statistical radio engineering. Radio and communications, Moscow, 656 p

83. Marple Jr SL (1990) Digital spectral analysis and its applications, Translation from English. Mir, Moscow, 584 p
84. Melnik YuA, Stogov GV (1973) Fundamentals of radio engineering and radio engineering devices. Soviet radio, Moscow, 368 p
85. Melnikov VM (1997) Meteorological informativeness of Doppler radars. In: Proceedings of the All-Russian Symposium «Radar studies of natural environments, issue 1. Military Engineering Space Academy, named after A.F. Mozhaisky, St. Petersburg, pp 165–172
86. Melnikov VM (1993) Information processing in Doppler weather radars. Foreign Radio Electron 4:35–42
87. Minkovich BM, Yakovlev VP (1969) Theory of antenna synthesis. Soviet radio, Moscow, 296 p
88. Mironov MA (2001) Estimation of the parameters of the autoregression model and the moving average, based on experimental data. Publishing house Radio engineering, no 10, pp 8–12
89. Mikhailutsa KT, Chernulich VV (1981) Digital signal processing device for incoherent weather radar. Questions of special radio electronics. Radar equipment series, no 20
90. Leonov AI, Vasenev VN, Gaidukov YuI et al (1979) Modeling in radar. In: Leonov AI (ed). Soviet radio, Moscow, 264 p
91. Manual on the production of flights in civil aviation of the USSR (Manual on the production of flights in civil aviation—85) (1985) Air transport, Moscow, 254 p
92. Nemov AV, Dobrn VV, Kuznetsova EV (2002) Joint use of super-resolving frequency estimates. News of Russian universities. Radioelectronics 2:85–92
93. Ostrovityanov RV, Basalov FA (1982) Statistical theory of radar of extended targets. Radio and communications, Moscow, 232 p
94. Problems of creation and application of mathematical models in aviation (1983). Nauka, Moscow (Series «Questions of cybernetics»)
95. Radar methods of Earth research. In: Melnik YuA. Soviet radio, Moscow, 264 p
96. Kondratenkov GS, Potekhin VA, Reutov AP, Feoktistov YuA (1983) Radar stations of the Earth survey. In: Kondratenkov GS (ed). Radio and communications, 272 p
97. Sarychev VA, Antsev GV (1992) Modes of operation of multifunctional airborne radar systems for civil purposes. Radioelectron Commun 4:3–8
98. Svistov VM (1977) Radar signals and their processing. Soviet radio, Moscow, 448 p
99. Handbook of climatic characteristics of the free atmosphere for individual stations of the Northern hemisphere. In: Guterman IG. Moscow
100. Tikhonov VI, Kulman NK (1975) Nonlinear filtering and quasi-coherent signal reception. Soviet radio, Moscow, 704 p
101. Uskov V (1987) Wind shear and its effect on landing. Civil Aviat 12:27–29
102. Feldman YuI, Mandurovsky IA (1988) Theory of fluctuations of location signals, reflected by distributed targets. Radio and communications, Moscow, 272 p
103. Khaykin S, Curry BU, Kesler SB. Spectral analysis of radar interfering reflections by the maximum entropy method
104. Shlyakhin VM (1987) Probabilistic models of Nerelev fluctuations of radar signals: a review. Radio Eng Electron 32(9):1793–1817
105. Vereshchagin AV, Zatuchny DA, Sinitsyn VA, Sinitsyn EA, Shatrakov YG (2020) Signal processing of airborne radar stations plane flight control in difficult meteoconditions. Springer Aerospace Technology, 218 p
106. Yurkov NK, Bukharov AYe, Zatuchny DA (2021) Signal polarization selection for aircraft radar control—models and Methods. Springer Aerospace Technology, 140 p

Printed in the United States
by Baker & Taylor Publisher Services